Commercial SUPERSONIC Technology

The Way Ahead

Committee on Breakthrough Technology for Commercial Supersonic Aircraft
Aeronautics and Space Engineering Board
Division on Engineering and Physical Sciences
National Research Council

NATIONAL ACADEMY PRESS
Washington, D.C.

NOTICE: The project that is the subject of this report was approved by the Governing Board of the National Research Council, whose members are drawn from the councils of the National Academy of Sciences, the National Academy of Engineering, and the Institute of Medicine. The members of the committee responsible for the report were chosen for their special competences and with regard for appropriate balance.

This project was supported by the National Aeronautics and Space Administration under contract No. NASW 99037. Any opinions, findings, conclusions, or recommendations expressed in this publication are those of the authors and do not necessarily reflect the views of the National Aeronautics and Space Administration.

International Standard Book Number: 0-309-08277-3

Available in limited supply from Aeronautics and Space Engineering Board, HA 292, 2101 Constitution Avenue, N.W., Washington, DC 20418, (202) 334-2855, <*www.nationalacademies.org/cets/asebhome.nsf*>.

Additional copies available for sale from National Academy Press, 2101 Constitution Avenue, N.W., Box 285, Washington, DC 20055, 1-800-624-6242 or (202) 334-3313 (in the Washington metropolitan area), <*www.nap.edu*>.

Printed in the United States of America

Cover: Bird illustrations designed by Antony Jameson, Stanford University.

THE NATIONAL ACADEMIES

National Academy of Sciences
National Academy of Engineering
Institute of Medicine
National Research Council

The **National Academy of Sciences** is a private, nonprofit, self-perpetuating society of distinguished scholars engaged in scientific and engineering research, dedicated to the furtherance of science and technology and to their use for the general welfare. Upon the authority of the charter granted to it by the Congress in 1863, the Academy has a mandate that requires it to advise the federal government on scientific and technical matters. Dr. Bruce M. Alberts is president of the National Academy of Sciences.

The **National Academy of Engineering** was established in 1964, under the charter of the National Academy of Sciences, as a parallel organization of outstanding engineers. It is autonomous in its administration and in the selection of its members, sharing with the National Academy of Sciences the responsibility for advising the federal government. The National Academy of Engineering also sponsors engineering programs aimed at meeting national needs, encourages education and research, and recognizes the superior achievements of engineers. Dr. Wm. A. Wulf is president of the National Academy of Engineering.

The **Institute of Medicine** was established in 1970 by the National Academy of Sciences to secure the services of eminent members of appropriate professions in the examination of policy matters pertaining to the health of the public. The Institute acts under the responsibility given to the National Academy of Sciences by its congressional charter to be an adviser to the federal government and, upon its own initiative, to identify issues of medical care, research, and education. Dr. Kenneth I. Shine is president of the Institute of Medicine.

The **National Research Council** was organized by the National Academy of Sciences in 1916 to associate the broad community of science and technology with the Academy's purposes of furthering knowledge and advising the federal government. Functioning in accordance with general policies determined by the Academy, the Council has become the principal operating agency of both the National Academy of Sciences and the National Academy of Engineering in providing services to the government, the public, and the scientific and engineering communities. The Council is administered jointly by both Academies and the Institute of Medicine. Dr. Bruce M. Alberts and Dr. Wm. A. Wulf are chairman and vice chairman, respectively, of the National Research Council.

COMMITTEE ON BREAKTHROUGH TECHNOLOGY FOR COMMERCIAL SUPERSONIC AIRCRAFT

DIANNE S. WILEY, *Chair*, The Boeing Company, Huntington Beach, California
H. LEE BEACH, JR., Christopher Newport University, Newport News, Virginia
JAMES A. (MICKY) BLACKWELL, Lockheed Martin (retired), Marietta, Georgia
EUGENE E. COVERT, Massachusetts Institute of Technology, Cambridge
DONALD M. DIX, Department of Defense (retired), McLean, Virginia
WILLARD DODDS, GE Aircraft Engines, Cincinnati, Ohio
ILAN KROO, Stanford University, Stanford, California
DOMENIC J. MAGLIERI, Eagle Aeronautics, Inc., Hampton, Virginia
MATTHEW MILLER, The Boeing Company, Seattle, Washington
DORA E. MUSIELAK, University of Texas, Arlington
DAVID K. SCHMIDT, University of Colorado, Colorado Springs
MICHAEL WINSLOW, Honeywell, Dayton, Ohio
BILL G.W. YEE, Pratt & Whitney (retired), West Palm Beach, Florida

Liaison from the Aeronautics and Space Engineering Board

ROBERT C. GOETZ, Lockheed Martin Skunk Works (retired), Santa Clara, California

Staff

PRABHAKAR MISRA, Study Coordinator
ALAN ANGLEMAN, Senior Program Officer
MARY LOU AQUILO, Senior Project Assistant
BRIDGET EDMONDS, Senior Project Assistant
GEORGE LEVIN, Director, Aeronautics and Space Engineering Board
MARVIN WEEKS, Senior Administrative Assistant

AERONAUTICS AND SPACE ENGINEERING BOARD

Staff

Preface

Affordable, reliable, and safe air transportation is important to quality of life and economic growth. Civil aviation has become an essential mode of transportation nationally and globally. If the United States intends to maintain supremacy in the commercial aerospace sector, it has to take a long-term perspective and channel adequate resources into research and technology development. In fact, this task is one of the legislatively established objectives of the National Aeronautics and Space Administration (NASA), but it will not be achieved without a vigorous aeronautics program that is relevant to the development of advanced commercial aircraft, including supersonic aircraft.

NASA's Aerospace Technology Enterprise has established 10 technology goals, one of which is to cut in half the time it takes to travel from the United States to the Far East and Europe. Achieving this objective will require new technology to improve the performance and affordability of supersonic aircraft while meeting public expectations related to safety, noise, and engine emissions. Advanced research and technology are also needed to establish the feasibility of reducing sonic boom sufficiently to allow sustained supersonic commercial flight over land—a capability that would greatly enhance the utility and economic viability of supersonic aircraft.

The National Research Council (NRC) was commissioned by NASA to conduct an 18-month study to identify breakthrough technologies for overcoming key barriers to the development of an environmentally acceptable and economically viable commercial supersonic aircraft. The NRC subsequently established the Committee on Breakthrough Technology for Commercial Supersonic Aircraft. The study committee met four times between October 2000 and March 2001 and also had other ancillary visits and teleconferences to collect relevant information, identify and assess alternative technologies, and generate a list of appropriate findings, conclusions, and recommendations. As detailed herein, the committee concluded that an economically viable supersonic aircraft will require new focused efforts in several areas, as well as continued development of technology on a broad front. Furthermore, NASA must advance key technologies to a technology readiness level (TRL) high enough (i.e., a TRL of 6, as defined by NASA) to facilitate the handoff of research results to the aerospace industry for commercial development. The committee concluded that maturation of key technologies could enable operational deployment of an environmentally acceptable, economically viable commercial supersonic aircraft with a cruise speed of less than approximately Mach 2 in 25 years or less—perhaps a lot less, with an aggressive technology development program focused on smaller supersonic aircraft, because goals in many critical areas would be easier to achieve with smaller aircraft. However, it may take longer to overcome the more difficult technological and environmental challenges associated with building a large commercial supersonic aircraft with a cruise speed in excess of approximately Mach 2.

This study benefited from a high level of public interest. Many individuals from interested organizations attended the committee's information-gathering meetings, which included opportunities for public input. This broad participation made an important contribution to the committee's deliberations, and the committee is indebted to everyone who gave of their time and talent at the meetings.

This report has been reviewed in draft form by individuals chosen for their diverse perspectives and technical expertise, in accordance with procedures approved by the NRC's Report Review Committee. The purpose of this independent review is to provide candid and critical comments that will assist the institution in making its published report as sound as possible and to ensure that the report meets institutional standards for objectivity, evidence, and responsiveness to the study charge. The review comments and draft manuscript remain confidential to protect the integrity of the deliberative process. We wish to thank the following individuals for their review of this report:

Linden Blue, General Atomics,
Michael Hudson, Rolls-Royce North America,
Antony Jameson, Stanford University,
Ira Kuhn, Directed Technologies, Inc.,
Kenneth Plotkin, Wyle Laboratories, and
William Sirignano, University of California, Irvine

Although the reviewers listed above have provided many constructive comments and suggestions, they were not asked to endorse the conclusions or recommendations nor did they see the final draft of the report before its release. The review of this report was overseen by Raymond S. Colladay, RC Space Enterprises, Inc. Appointed by the National Research Council, he was responsible for making certain that an independent examination of this report was carried out in accordance with institutional procedures and that all review comments were carefully considered. Responsibility for the final content of this report rests entirely with the authoring committee and the institution.

Dianne S. Wiley, *Chair*
Committee on Breakthrough Technology
for Commercial Supersonic Aircraft

Contents

APPENDIXES

Tables, Figures, and Box

TABLES

FIGURES

BOX

WORKING TOGETHER

We shape our self
to fit this world

and by the world
are shaped again.

The visible
and the invisible

working together
in common cause,

to produce
the miraculous.

I am thinking of the way
the intangible air

passed at speed
round a shaped wing

easily
holds our weight.

So may we, in this life
trust

to those elements
we have yet to see

or imagine,
and look for the true

shape of our own self
by forming it well

to the great
intangibles about us.

—David Whyte

Written for the presentation of the Collier Trophy to the Boeing Company, 1996.

Executive Summary

High-speed flight is a major technological challenge for both commercial and business aviation. As a first step in revitalizing efforts by the National Aeronautics and Space Administration (NASA) to achieve the technology objective of high-speed air travel, NASA requested the National Research Council (NRC) to conduct a study that would identify approaches for achieving breakthroughs in research and technology for commercial supersonic aircraft. This report documents the results of that effort. It describes technical areas where ongoing work should be continued and new focused research initiated to enable operational deployment of an environmentally acceptable, economically viable commercial aircraft capable of sustained supersonic flight, including flight over land, at speeds up to approximately Mach 2 in the next 25 years or less. In particular, sonic boom is the major barrier to development of supersonic business jets (SBJs) and a major, but not the only, barrier to the development of supersonic transports with overland capability. While NASA should have its eye on the grand prize—supersonic commercial transports—it is still quite appropriate for NASA to conduct sonic boom research, even when related to SBJs. The report also identifies other critical areas where technology development is needed to support the development of commercial supersonic aircraft with cruise speeds beyond Mach 2.

STUDY OBJECTIVE AND APPROACH

The objective of this study was to leverage the results of NASA's High Speed Research (HSR) Program, other research, and related studies to identify breakthrough technologies for overcoming key barriers to the development of an environmentally acceptable, economically viable commercial supersonic aircraft with minimal sonic boom (to enable supersonic flight over land). The scope of the study included both small aircraft (that is, SBJs) and large transports. The focus of the study was on high-risk, high-payoff technologies where NASA-supported research could make a difference over the next 25 years. The committee did not focus on any specific vehicle configuration, market segment, or technology readiness level (TRL), although it believes that, to have practical value, government-funded research should advance to a TRL of at least 6 before industry can be expected to incorporate new technologies into commercial aviation products.[1]

The committee considered which of two options it would focus on:

1. revolutionary new types of aircraft that are fundamentally different from existing aircraft
2. vehicles that more closely resemble existing aircraft

Both options require breakthrough technologies, but developing a new vehicle concept into an operational commercial supersonic aircraft would take decades of research and development to satisfy aircraft performance, economic, safety, certification, and environmental requirements. Focusing on revolutionary vehicle concepts to develop commercial supersonic aircraft would probably give NASA a research program that does little or nothing to enable operational deployment of commercial supersonic aircraft within the next 25 years. Because that is the time frame of interest for this study, the committee concentrated on identifying areas where breakthroughs and focused investments are most likely to achieve the ultimate objective of sustained commercial supersonic flight, including flight over land, in a more timely fashion.

As the study proceeded it became clear that the statement of task presented procedural challenges. Breakthrough tech-

[1]NASA rates the level of technology readiness on a scale of 1 to 9. TRL 6 has been achieved when a system or subsystem model or prototype has been demonstrated in a relevant environment.

nologies are likely to be guarded as proprietary and competition-sensitive and are not available to groups such as NRC committees, which work in a public forum. This was true for much of the results of the HSR Program, which is not yet public information, as well as details of ongoing work, such as the Quiet Supersonic Platform Program funded by the Defense Advanced Research Projects Agency (DARPA). Because of its limited access to detailed technical data, the committee was reluctant to accept the role of recommending which individual technical concepts and approaches, many of them in the early stages of development, should be funded. Fortunately, it was not necessary for the committee to do so. As demonstrated by the Quiet Supersonic Platform Program, once the government identifies specific areas of interest and allocates development funding, industry and other research organizations are able and willing to provide detailed, innovative research proposals within the framework of a competitive government acquisition program, where competition-sensitive information is more likely to be protected. Therefore, with the concurrence of the study sponsor, the committee carried out the intent of the statement of task using the following approach:

1. Identify the technical barriers to sustained commercial flight, including flight over land.
2. Characterize the gap between the state of the art and the technology required to overcome each barrier.
3. Establish the feasibility of closing each gap by considering if at least one promising approach is available.
4. Identify what would have to be demonstrated to show that the gap has been closed.

TECHNOLOGY CHALLENGES

To provide a framework for analyzing the technology needs for commercial supersonic flight over the next 25 years, the committee defined a set of three notional supersonic vehicles:

- *Small.* An SBJ with about 8 to 15 passengers, a range of 4,000 to 5,000 nautical miles (NM), a cruise speed of about Mach 1.6 to 1.8, and sonic boom low enough to enable supersonic flights over both land and water.
- *Medium.* An overland supersonic commercial transport with about 100 to 200 passengers, a range of 4,000 to 5,000 NM, a cruise speed of about Mach 1.8 to 2.2, and sonic boom low enough to enable operations over both land and water.
- *Large.* A high-speed civil transport (HSCT) with about 300 passengers, a range of 5,000 to 6,000 NM, and a cruise speed of about Mach 2.0 to 2.4.

The committee also compared the technical challenges for commercial supersonic aircraft with the likely challenges for a military supersonic strike aircraft. A strike aircraft would have to overcome many of the same challenges as a commercial aircraft—for example, a high lift-to-drag ratio and acceptable takeoff and landing characteristics; efficient and durable engines; and advanced airframe materials and structures. A strike aircraft would need to meet these challenges while also meeting military requirements for stealth and weapons integration, but without necessarily meeting all the same environmental constraints.

For each class of aircraft, the committee used a combination of engineering judgment, historical trends, and simplified equations to identify key challenges and the research areas required to overcome those challenges. While noise and emissions are certainly major barriers to the development of an HSCT, significant advances in the traditional aeronautics engineering disciplines, such as structures, propulsion, and aerodynamics, are still required to close the business case and certificate new systems. Supersonic transports with overland capability (and military strike aircraft of comparable size) will require improvements in the four major factors related to economics (lift-to-drag ratio, air vehicle empty weight fraction, specific fuel consumption, and thrust-to-weight ratio) equivalent to about 10 percent over the present state of the art in each parameter, as well as additional advances related to the environment and certification. For SBJs, most parameters are already within the state of the art. HSCTs, on the other hand, will require significant advances, equivalent to about 15 percent for each of the four major economic parameters. Affordable supersonic flight is an exercise in integration: A viable commercial supersonic aircraft cannot be achieved until solutions to the individual technology challenges are brought together in one integrated airframe-engine design.

The committee also validated the importance of cruise speed as a key factor in determining the technological difficulty associated with development of a commercial supersonic aircraft. NASA's HSR Program, which ran from 1985 to 1999, envisioned an HSCT with 300 passengers and a cruise speed of Mach 2.4. In 1997 the NRC concluded that the focus on Mach 2.4 was too aggressive and probably not justified by the business analysis. The study concluded that an aircraft with a cruise speed of Mach 2.0 might have a net productivity similar to that of a Mach 2.4 aircraft and would have an easier time overcoming some of the most difficult economic, technological, and environmental challenges.

As cruise speed increases, the most efficient cruise altitude increases also, and the technical challenges to developing an economically viable and environmentally acceptable commercial supersonic aircraft increase significantly above approximately Mach 2. For aircraft with cruise speeds less than Mach 2, an NO_x emission index of 15 appears satisfactory, and water vapor emissions are unlikely to pose difficulties at the associated altitudes. Aircraft with cruise speeds in excess of Mach 2 will normally cruise in the stratosphere, where engine emissions have a greater potential to cause climate change and depletion of atmospheric ozone. At higher speeds, the NO_x emissions index may need to be as low as 5.

In addition, water vapor, which is benign in the lower atmosphere, may have significant, long-lasting effects in the stratosphere. Also, at higher speeds air friction creates higher temperatures. For cruise speeds above approximately Mach 2.2, new structural materials are needed to meet requirements for strength, weight, and affordability. The noise suppression problem also becomes more challenging as speed increases, because cruise efficiency requirements mandate the use of engines with lower bypass ratios and, accordingly, higher jet exhaust velocities and more noise. As a result, larger nozzles are required to meet community noise standards. Without advanced technology, the nozzles of aircraft with cruise speeds above Mach 2.2 will probably be too large and heavy to be economical.

The intensity of sonic booms increases with vehicle size, weight, and speed; developing a low-boom design suitable for supersonic cruise over land will be much harder for an HSCT than for smaller transports or an SBJ. Even so, the development of any economically viable commercial supersonic aircraft is far from trivial. High-risk investments are still required to develop and validate the design of a small supersonic aircraft with low sonic booms. Success in this endeavor, however, could support the eventual development of an HSCT with a low sonic boom by performing critical noise suppression experiments, testing public acceptance of sonic boom noise levels, and gathering critical data.

RECOMMENDATIONS FOR RESEARCH

The technology advances necessary to attain the economic and environmental goals will be easiest to achieve for an SBJ and most difficult for an HSCT. Surmounting the special technology challenges associated with cruise speeds greater than approximately Mach 2 will require long-term research. To achieve commercial supersonic flight by 2025, NASA's primary focus should be to support (1) new initiatives and (2) ongoing research that would enable supersonic flight at cruise speeds less than approximately Mach 2, where the technology challenges are more tractable in the near- to midterm. Specific areas of research identified by the committee appear in the findings and recommendations below and are also shown in Table ES-1.

Finding 1. An economically viable, environmentally acceptable supersonic commercial aircraft with a cruise speed of less than approximately Mach 2 requires continued development of technology on a broad front (see Finding 2). In addition, research in the following five areas of critical impor-

TABLE ES-1 Key Challenges to Developing a Commercial Supersonic Aircraft and Related Research Areas

Characteristics of Supersonic Aircraft: Key Challenges / Related Research Areas	Long-Life, Lightweight Airframe Materials and Structures	Long-Life, High-Temperature Airframe Materials	Airframe Configurations	Laminar Flow	Flight-Critical, Highly Augmented Automatic Control Systems	Operational Procedures and Limitations for Noise Abatement	Synthetic Vision	Aero-propulsive servo-elastic and Aircraft-pilot servo-elastic (APSE)	Design Tools and Methods	Certification Standards for New Technologies and Procedures	Cabin Depressurization Safety	Radiation Exposure Safety	Human Response to Sonic Booms	Propulsion Systems	Variable Cycle Engine	Lightweight Noise Suppression Systems	Long-Life, High-Temperature Engine Materials	Effects of Emissions (H_2O, NO_x, etc.) in the Stratosphere
Economics—range, payload, fuel burn, etc.																		
Low weight and low empty weight fraction	2						2								1			
Improved aerodynamic performance			1	1			2											
Highly integrated airframe/propulsion system									1									
Low thrust-specific fuel consumption															1			
Long life	2	3															2	
Environment																		
Benign effect on climate and atmospheric ozone																		3
Low noise during takeoff and landing[a]			2		2	2								2	1	3		
Low sonic boom			1							1			1					
Certification for commercial operations										2								
Acceptable handling and ride qualities					1			1	1									
Passenger and crew safety at high altitudes											2	2						

Airframe ← → Integrated Aircraft ← → Propulsion System

NOTE: 1, research area recommended by Finding 1; 2, research area recommended by Finding 2; 3, research area recommended by Finding 3.

tance could lead to important breakthroughs, but only if current research is augmented by new, focused efforts (or significant expansions of existing efforts):

- airframe configurations to reduce sonic boom intensity, especially with regard to the formation of shaped waves and the human response to shaped waves (to allow developing an acceptable regulatory standard)
- improved aerodynamic performance, which can be achieved through laminar flow and advanced airframe configurations (both conventional and unconventional)
- techniques for predicting and controlling aero-propulsive servo-elastic and aircraft-pilot servo-elastic (APSE) characteristics, including high-authority flight- and structural-mode control systems for limiting both types of APSE effects in flight and tools for defining acceptable handling and ride qualities
- automated, high-fidelity, multidisciplinary optimization tools and methods for design, integration, analysis, and testing of a highly integrated, actively controlled airframe-propulsion system
- variable cycle engines for low thrust-specific fuel consumption, high thrust-to-weight ratio, and low noise

Recommendation 1. NASA should focus new initiatives in supersonic technology development in the areas identified in Finding 1 as they apply to aircraft with cruise speeds of less than approximately Mach 2. Such initiatives should be coordinated with similar efforts supported by other federal agencies (e.g., the DARPA Quiet Supersonic Platform Program).

In addition to focused research initiatives on the technologies listed in Finding 1, the development of economically viable, environmentally acceptable commercial supersonic aircraft also requires continued advances in other areas. A broad range of technical activity must be supported to ensure that other key technologies are mature enough to convince industry to take on development and to meet designer's needs at the time a supersonic transport is laid down. Although advances in these additional technologies (see Finding 2) may not be necessary to develop an SBJ, they are critical to the ultimate goal of developing a large commercial supersonic transport.

Finding 2. An economically viable, environmentally acceptable commercial supersonic transport with a cruise speed of less than approximately Mach 2 requires continued advances in many areas, particularly the following:

- airframe materials and structures for lower empty weight fractions and long life, including accelerated methods for collecting long-term aging data and the effects of scaling on the validity of thermomechanical tests
- engine materials for long life at high temperatures, including combustor liner materials and coatings, turbine airfoil alloys and coatings, high-temperature alloys for compressor and turbine disks, and turbine and compressor seals
- aerodynamic and propulsion systems with low noise during takeoff and landing
- cockpit displays that incorporate enhanced vision systems
- flight control systems and operational procedures for noise abatement during takeoff and landing
- certification standards that encompass all new technologies and operational procedures to be used with commercial supersonic aircraft
- approaches for mitigating safety hazards associated with cabin depressurization at altitudes above about 40,000 ft
- approaches for mitigating safety hazards that may be associated with long-term exposure to radiation at altitudes above about 45,000 ft (updating the Federal Aviation Administration's advisory circular on radiation exposure, AC 120-52, to address supersonic aircraft would be a worthwhile first step)

Recommendation 2. For the technologies listed in Finding 2, NASA should allocate most of the available resources on goals and objectives relevant to aircraft with cruise speeds of less than approximately Mach 2. NASA should focus remaining resources on the areas listed in Finding 3 (i.e., the highest risk areas for cruise speeds greater than approximately Mach 2). Again, NASA activities should be coordinated with similar efforts supported by other federal agencies.

Conclusion 1. Research and technology development in the areas listed in Findings 1 and 2 could probably enable operational deployment of environmentally acceptable, economically viable commercial supersonic aircraft in 25 years or less—perhaps a lot less, with an aggressive technology development program for aircraft with cruise speeds less than approximately Mach 2.

Finding 3. An economically viable supersonic commercial aircraft with a cruise speed in excess of approximately Mach 2 would require research and technology development in all of the areas cited in Findings 1 and 2. In addition, significant technology development would be needed to overcome the following barriers:

- climate effects and depletion of atmospheric ozone caused by emissions of water vapor and other combustion by-products in the stratosphere
- high temperatures experienced for extended periods of time by airframe materials, including resins, adhesives, coatings, and fuel tank sealants

- noise suppression at acceptable propulsion system weight

Conclusion 2. Candidate technologies for overcoming environmental barriers to a commercial supersonic aircraft with a cruise speed in excess of approximately Mach 2 are unlikely to mature enough to enable operational deployment of an environmentally acceptable, economically viable Mach 2+ commercial supersonic aircraft during the next 25 years.

The ultimate importance of the commercial supersonic aircraft to the U.S. air transportation system is expounded in the long-range technology plans and visions of NASA, the Department of Transportation, and the National Science and Technology Council. Fulfilling these visions of the future will require a long-term investment strategy that looks beyond the short-term economic factors that drive much industry-funded research. The importance of a long-term view is especially important with breakthrough technologies. Unfortunately, both government and industry are reluctant to make the long-term investments necessary to mature expensive, high-risk technologies. In particular, at a time when manufacturers require TRLs of 6 or higher to embrace complex new technologies in safety-critical aeronautics applications, NASA appears to be changing its technology investment strategy so that it reaches TRLs of only 3 or 4. The likely result is a technology maturation gap that could jeopardize U.S. leadership in aerospace technology. To avoid this result—that is, to allow promising technologies to make the transition from the laboratory to the marketplace—NASA must invest enough to achieve TRL 6. With past programs, such as the HSR Program, NASA adopted TRL 6 as the appropriate goal for commercial supersonic research, and the committee is concerned that NASA's less ambitious goals for much of its ongoing aeronautics research is driven more by the need to curtail aeronautics research because of reduced funding than by an objective assessment of what it will take to achieve the government's programmatic goals.

Recommendation 3. NASA and other federal agencies should advance the technologies listed in Findings 1 and 2 and Recommendations 1 and 2 to technology readiness level 6 to make it reasonably likely that they will lead to the development of a commercial product.

In summary, the committee identified no insurmountable obstacles to the development of commercial supersonic aircraft and believes that a properly focused research effort by NASA could develop technological solutions to the key problems identified in Finding 1, thereby enabling a successful commercial development program by industry in the relatively near term, especially for aircraft with a cruise speed of less than Mach 2.0. Without continued effort, however, an economically viable, environmentally acceptable commercial supersonic aircraft is likely to languish. National indifference to supersonic technology development would jeopardize longstanding U.S. supremacy in the aviation business segment and significantly harm the nation's economy. The United States is not the only sponsor of supersonic technology development, and once a commercial supersonic aircraft is developed, users in the United States and other countries will purchase it, regardless of where it is manufactured.

1

Introduction

High-speed flight is a major technological challenge for commercial and business aviation. To help meet this challenge, the National Aeronautics and Space Administration (NASA) asked the National Research Council (NRC) to conduct a study that would identify approaches for achieving breakthroughs in supersonic research and technology (R&T). This report documents the results of that effort.

The report is organized into five chapters. This introduction describes the study process and the committee's understanding of NASA's expectations for the study. It also provides background information to set the context for the report's key findings, conclusions, and recommendations. Chapter 2 briefly reviews how the business case for commercial supersonic aircraft has changed since NASA's High Speed Research (HSR) Program was cancelled in 1999 and discusses notional vehicles around which technology challenges and research time frames may be grouped. Chapter 3 describes five areas for new focused research to enable industry production by the year 2025 of an environmentally acceptable, economically viable commercial aircraft capable of sustained supersonic flight, including flight over land, at speeds less than approximately Mach 2. Chapter 4 identifies additional critical areas where continued technology development is needed. Chapter 5 presents findings, conclusions, and recommendations that follow from and summarize the earlier discussions. Biographies of committee members are presented in Appendix A. Participants in meetings of the full committee and key subcommittee meetings are listed in Appendix B. Acronyms and abbreviations used in the report are listed in Appendix C.

STATEMENT OF TASK AND STUDY APPROACH

The statement of task for the study was as follows:

- Based on the results of NASA's HSR Program, other research, and related studies, identify key customer and design requirements that cannot be satisfied by currently available technology or by adapting technology that is likely to be developed for other applications.
- Identify breakthrough technologies that may be able to satisfy the high-risk requirements identified above.
- Prepare a report that qualitatively assesses the most promising breakthrough technologies in terms of their potential value and risk.

The scope of the study included both small aircraft (i.e., supersonic business jets [SBJs]) and large transports. However, the committee was directed not to develop a comprehensive plan addressing all of the R&T needs for a commercial supersonic aircraft. Therefore, this report does not address many important aircraft systems, such as onboard power systems, where government research is not critical to the development of future commercial supersonic aircraft.

The study focused on high-risk, high-payoff technologies where NASA-supported research could make a difference over the next 25 years. It did not focus on any specific vehicle configuration, market segment, or technology readiness level (TRL), although the committee believes that, to have practical value, government-funded research should achieve a TRL range of at least 6 before industry can be expected to incorporate new technologies into commercial aviation products (see Figure 1-1).

The committee conducted hearings and fact-finding meetings to identify technical barriers and promising breakthrough technologies. However, as the study proceeded it became clear that the statement of task presented difficulties with respect to its use of the phrase "breakthrough technology." Truly breakthrough technologies are likely to be guarded as proprietary and competition sensitive and, as such, are not available to groups such as the NRC committees, which work in a public forum. Thus, even though many committee members knew about potential breakthrough

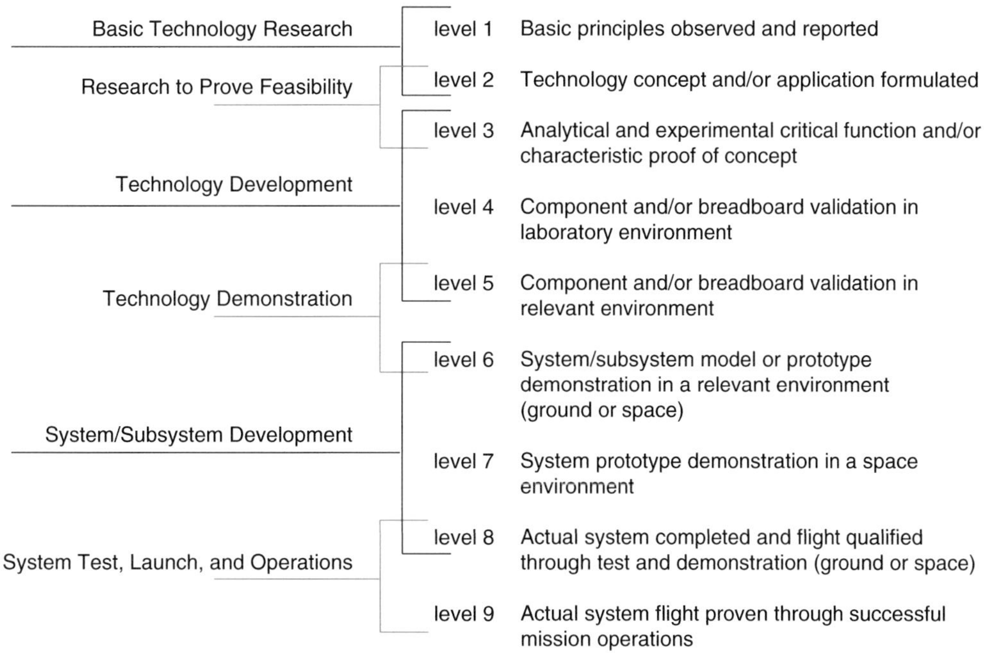

FIGURE 1-1 NASA technology readiness levels. SOURCE: NASA.

technologies, the arrangements under which such knowledge was obtained prevented it from being disclosed or included in this report. This included much of the results of the HSR Program, which are classified as limited exclusive rights data (LERD), details of ongoing work supported by the Defense Advanced Research Projects Agency (DARPA) Quiet Supersonic Platform (QSP) Program, and information collected through work as employees of or consultants to private companies.

Because of its limited insight into specific technological solutions to the problems of commercial supersonic flight, the committee was reluctant to accept the role of recommending which individual technical concepts and approaches, many of them in early stages of development, should be funded. For example, one possible breakthrough approach would be the application of "lightcraft technologies," which are being developed by Lightcraft Technologies, Inc., and Rensselaer Polytechnic University, supported in part by the U.S. Air Force and NASA (David, 2000). This technology suite transcends the conventional notion of a supersonic aircraft by conceptualizing aircraft powered by ground-based lasers. Even if the remaining technical issues are ultimately resolved, however, commercialization as a passenger vehicle will not be achieved in the 25-year time frame that is the focus of this report. Another interesting technology would inject water or other fluid in the engine exhaust to mitigate takeoff noise, allowing the use of smaller exhaust nozzles. The committee also became aware of a proposal for boomless supersonic flight, but the concept has yet to be submitted to peer review and available data are insufficient to allow judging its viability.

Rather than attempt to identify, evaluate, and compare the merits of individual breakthrough technologies, the committee concentrated instead on identifying problems and areas where breakthroughs and focused investment are needed to achieve the ultimate objective of sustained commercial supersonic flight, including flight over land. As demonstrated by DARPA's QSP Program, if the government decides to address a particular problem—such as the challenge of supersonic flight with very low (or no) sonic boom—R&T solicitations would attract proposals for many diverse solutions, which can then be evaluated as part of the solicitation process. Furthermore, NASA has especially good insight into the work of the QSP Program because the program manager is a NASA civil servant on temporary assignment at DARPA. In light of the concerns expressed above, and with the concurrence of the study sponsor, the committee used the following approach to guide it in carrying out the intent of the statement of task:

1. Identify the technical barriers to sustained commercial supersonic flight, including flight over land.
2. Characterize the gap between the state of the art and the technology required to overcome each barrier.
3. Establish the feasibility of closing each gap by considering if at least one promising approach is available.
4. Identify what would have to be demonstrated to show that the gap has been closed.

BACKGROUND

Commercial supersonic aircraft have been the object of many development programs. The first major effort by the United States, the Supersonic Transport (SST) Program, was terminated in the early 1970s prior to completion. France and Great Britain, however, were able to complete development of the Concorde, which has shown both the possibilities and problems of commercial supersonic aircraft. NASA's HSR Program began in 1985 with the objective "to establish the technology foundation by 2002 to support the U.S. transport industry's decision for a 2006 production of an environmentally acceptable, economically viable, 300-passenger, 5,000 nautical mile, Mach 2.4 aircraft" (NRC, 1997). The first flight of a commercial supersonic aircraft was envisioned around 2010, with the first production aircraft to be operational around 2013. As shown in Table 1-1, NASA expected that program goals would require a large investment.

The HSR Program was conducted in two phases. Phase 1, from 1985 to 1990, focused on environmental compatibility issues. Phase 2, from 1990 until 1999, focused on development of airframe and propulsion technologies. A proposed but never implemented Phase 2A would have focused on affordable manufacturing technologies starting in 1997. The HSR Program was canceled in 1999, largely because NASA was unwilling to continue the program without a stronger commitment from industry to the commercial development of a 300-passenger, Mach-2.4 aircraft. As a result, NASA's supersonic R&T efforts were greatly curtailed.

Regulatory standards are a key factor in determining the viability of commercial supersonic aircraft. Because environmental standards are becoming more strict, delays in developing a supersonic commercial aircraft mean that when one is built, it might have to meet more stringent standards than originally anticipated by the HSR Program. Just before the program was closed out, HSR managers solicited new ideas on how to meet more stringent environmental standards. This admittedly hasty exercise was intended to provide a technology development roadmap in anticipation of future efforts to develop supersonic technology.

The 1997 NRC report that assessed NASA's HSR Program emphasized the importance of relating aircraft performance requirements to the needs of customers—that is, passengers, airlines, manufacturers, and society. It concluded that the focus on Mach 2.4 was too aggressive and probably not justified by the business analysis. The report prioritized aircraft performance requirements based, in part, on their contribution to meeting customer needs and the level of risk and concluded that more advanced R&T is needed to establish the feasibility of sustained supersonic commercial flight. It presented a plan to achieve a flight demonstration aircraft that would fully validate aerodynamic, propulsion, and flight control technologies.

The United States is not the only sponsor of supersonic technology. The governments of France, Japan, Russia, and the United Kingdom are also sponsoring development of supersonic technology with commercial applications, although none has embarked on a formal program to produce a new commercial supersonic aircraft. The development of a commercial supersonic transport that can meet international environment standards and compete successfully with subsonic transports may be a larger effort than the industry of any single nation, including the United States, might wish to undertake. As with many such innovations, the first manufacturer to market will have the potential to dominate the market. If only one commercial supersonic transport is available, airlines from the United States and other countries will purchase it regardless of where it is manufactured. A small supersonic jet could be developed by a single aircraft manufacturer and might lead to important technological innovations.

The ultimate importance of commercial supersonic aircraft to the U.S. air transportation system is set forth in long-range technology plans and visions promulgated by NASA (NASA, 1998), the Department of Transportation (DOT, 1999), and the National Science and Technology Council (NSTC, 2000). Fulfilling these visions of the future will re-

TABLE 1-1 Total NASA Funding for the HSR Program from Program Inception in FY 1990 Through Planned Completion in FY 2002 (millions of dollars)

Program Element	FY 1990–1996	FY 1997	FY 1998-2002	Total
Propulsion	459.3	114.1	312.5	885.9
Airframe	322.6	110.5	286.8	719.9
Systems Integration	144.0	29.7	107.0	280.7
Total	925.9	254.3	706.3	1,886.5

SOURCE: NRC (1997).

quire a long-term investment strategy that looks beyond the shorter-term economic factors that drive much industry-funded research. A long-term view is especially important when it comes to breakthrough technologies. By nature, these technologies require long-term investments to increase their maturity and reduce their risk enough to convince the aerospace manufacturing industry to turn them into commercial products that the Federal Aviation Administration (FAA) will certify and airlines (or other users) will buy. Unfortunately, both government and industry are reluctant to make the long-term investments necessary to advance expensive, high-risk technologies. In particular, at a time when manufacturers require a TRL 6 or higher to embrace complex new technologies in safety-critical aeronautics applications, NASA appears to be redirecting its technology investment strategy to achieve an end point of TRL 3 or 4. The likely result is a technology maturation gap that could jeopardize U.S. leadership in aerospace technology. To avoid this consequence—that is, to allow promising technologies to make the transition from the laboratory to the marketplace—NASA must accept the challenge of investing enough to achieve TRL 6.

REFERENCES

David, L. 2000. "Laser-Boosted Rocket Sets Altitude Record." *Space.com*. November 3. Available online at <http://www.space.com/businesstechnology/technology/laser_craft_001103.html>. Accessed on September 13, 2001.

DOT (Department of Transportation). 1999. Effective Global Transportation in the Twenty-First Century: A Vision Document. U.S. Department of Transportation Research and Technology Coordinating Council "One DOT" Working Group on Enabling Research, September. Available online at <http://www.volpe.dot.gov/infosrc/strtplns/dot/glbtrn21/index.html#toc>. Accessed on September 13, 2001.

NASA (National Aeronautics and Space Administration). 1998. NASA Technology Plan, December. Available online at <http://technologyplan.nasa.gov/default.cfm?id=1.0>. Accessed on September 13, 2001.

NRC (National Research Council). 1997. U.S. Supersonic Commercial Aircraft. Available online at <http://www.nap.edu/catalog/5848.html>. Accessed on September 13, 2001.

NSTC (National Science and Technology Council) Committee on Technology, Subcommittee on Transportation Research and Development. 2000. National Transportation Strategic Research Plan. May. Available online at <http://www.volpe.dot.gov/infosrc/strtplns/nstc/srplan00/index.html>. Accessed on September 13, 2001.

2

Technology Challenges

To identify the barriers to developing supersonic aircraft in the next 25 years for which advances in technology are needed, it is first necessary to identify the key customer and design requirements for useful aircraft and translate them into technology needs. The approach used here assesses the capabilities desired, defines notional vehicles providing such capabilities, and analyzes the technology needs of these vehicles.

CUSTOMER REQUIREMENTS

Determining the level of technology required to meet customer requirements, ensure economic performance, and comply with environmental regulations requires some consideration of potential products. The process for creating product specifications consistent with satisfactory levels of performance in the above areas is complex, and the committee made no detailed attempts in this direction. However, committee members did interview domestic and international representatives, including business jet manufacturers and operators, large aircraft manufacturers, airlines, and the U.S. Air Force (USAF). The results of these interviews, combined with other information collected by the committee (summarized below), provided general guidance regarding the probable future of commercial and military supersonic aircraft.

Supersonic Business Jet

Prospective manufacturers, with encouragement from business jet fleet operators, view the SBJ as a near-term prospect, with a payload of perhaps 8 to 15 passengers, a cruise speed of approximately Mach 1.6 to Mach 1.8, and a range of 4,000 to 5,000 nautical miles (NM). Supersonic flight over land is essential for this class of vehicles, and the potential market is estimated to be at least 200 aircraft over a 10-year period. Economic factors are not nearly as limiting for business jets as for a commercial transport; prospective manufacturers believe the market will support paying about twice as much for a supersonic aircraft that can cruise at twice the speed of current subsonic business jets.

The consensus view of prospective manufacturers is that the key technology barrier for this class of aircraft is the elimination, or reduction to acceptable levels, of sonic boom for flight over land. Another barrier is the need for an engine that can operate for 2,000 hours between major overhauls. Significant advances in aerodynamic performance may also be required to achieve the desired aircraft range if the solution to the sonic boom problem imposes aircraft weight or performance penalties.

Military Strike Aircraft

Interviews with USAF personnel indicated that a long-range strike aircraft with sustained supersonic flight would provide significant potential for improving USAF war-fighting capabilities through substantially increased sortie rates and rapid response times. Desired characteristics are a payload of 20,000 to 40,000 pounds, a cruise speed of Mach 1.6 to 3.0, and a range of 4,000 to 6,000 NM. The precise Mach number is strongly influenced by optimization of total payload delivery rate (lb-NM/hour) for a given fleet cost. Long range is needed to reduce the need for forward basing and minimize the impact of overflight restrictions by countries not involved in a particular conflict. However, military aircraft would not necessarily need to meet all of the environmental constraints that apply to commercial aircraft.

The key technical challenges for a supersonic strike aircraft are as follows: development of a stealthy configuration with a high lift-to-drag ratio (L/D) and acceptable take-off and landing characteristics; efficient and durable engines; propulsion-airframe integration; advanced airframe materials and structures; weapons integration; and producibility.

Supersonic Commercial Transport

The desired payload for a high-speed civil transport (HSCT) would be about 300 passengers, and the desired range would be at least 4,500 NM, with 5,000 to 6,000 NM preferred. Sonic boom reduction would not be required to provide trans-oceanic service. The minimum expected cruise speed is approximately Mach 2.0, based in part on earlier studies of aircraft utilization on long international routes. A cruise Mach number close to 2 might be acceptable to users (i.e., airlines) and would also mitigate some of the technological challenges and environmental concerns associated with Mach numbers of 2.4 and greater.

A large transport, such as an HSCT, that can fly at Mach 2 or greater with little or no sonic boom, a capability necessary for overland operations, is not viewed as technologically feasible. One alternative for providing supersonic airline service over land would be to develop a large commercial supersonic aircraft with a cruise speed close enough to Mach 1 to avoid creating a sonic boom that would propagate to the ground. Boeing is currently conducting design studies for such an aircraft, which could probably be developed without government research into the breakthrough technologies that are the subject of this report.

A second alternative for improving the economics of a large supersonic transport would be to design it for flight at a high cruise speed over water and a lower (but still supersonic) cruise speed over land. This scenario, however, raises questions about performance efficiency during long flight segments at off-design speeds. Early in the HSR Program, for example, Boeing explored the option of aerodynamically shaping a Mach 2.4 aircraft to produce a low-level boom for overland flight at Mach 1.6 to Mach 1.8. This effort was dropped when it became clear that the necessary design modifications would significantly degrade performance at Mach 2.4 and reduce the overall economic viability of the aircraft. It might be worthwhile, however, to reexamine this area using advanced technologies and lower cruise speeds (for example, Mach 2.0 over water/Mach 1.2 over land instead of Mach 2.4 over water/Mach 1.6 over land).

A third alternative would be to build a commercial supersonic transport with a payload capability similar to that of a military strike aircraft (equivalent to 100 passengers), a range capability similar to that of an SBJ, and a cruise speed of approximately Mach 2. Compared to an HSCT, the reduced size and potentially lower speed would make it much more feasible to develop the technologies necessary to reduce the sonic boom enough to permit overland operations. This transport would be capable of transcontinental service as well as transoceanic service directly to and from noncoastal cities.

A 1997 study by the NRC reviewed demand studies that predicted a market size on the order of 1,000 HSCTs, assuming that targeted levels of cost, performance, and environmental impacts can be achieved (NRC, 1997). The demand studies also assumed that sonic booms would prevent these aircraft from flying at supersonic speeds over land and that they could be operated profitably with a ticket surcharge of about 10 percent (relative to the price of travel on subsonic aircraft) for coach class travel and a surcharge of 30 percent for business and first class travel. However, generalizations in the assumptions on which some of the demand studies were based may have caused the studies to overstate the projected market size. The 1997 NRC study concluded that turnaround times seemed to be unrealistically low and that an aircraft with a cruise speed of Mach 2.0 might have a productivity similar to that of a Mach 2.4 aircraft (NRC, 1997). A cruise speed of Mach 2.2 or less would also be a more tractable goal than Mach 2.4 for the 25-year period of interest to this study.

VEHICLE CHARACTERISTICS AND ECONOMIC GOALS

The above-mentioned views of customer requirements form the basis for defining three notional commercial supersonic aircraft: an SBJ, an overland supersonic commercial transport, and an HSCT. These generic aircraft were selected as being representative of the complete spectrum of supersonic aircraft likely to be developed in the foreseeable future. For example, the committee determined that the speed, range, and payload characteristics of an overland supersonic transport would be similar to those of a nominal supersonic strike aircraft. The purpose of these notional vehicles is to help assess the need for advances in the technological state of the art; they are not intended to endorse any particular product or replace the need for detailed design and market studies to validate the vehicle performance specifications prior to advanced product development.

Implicit in aircraft design specifications are costs—development, production, operations, and maintenance costs—that are considered affordable. Each of these costs depends on many factors. The 1997 NRC report identified 22 factors that impact vehicle affordability. For the purpose of this study, takeoff gross weight (TOGW) was used as one readily measurable indicator of aircraft cost for rapid assessment of configuration trade studies. Given this simplification, affordability is tied to the ratio of payload weight to TOGW that is considered to be economically viable (or militarily cost-effective) in the applications of interest identified above.

For each of the three notional commercial supersonic aircraft, the committee used a combination of engineering judgment, historical trends, and simplified equations to identify vehicle characteristics and the technology goals that must be achieved to satisfy requirements for an environmentally acceptable and economically viable aircraft (see Table 2-1 and Figure 2-1).

Overland transport aircraft (and comparably sized military strike aircraft) will require improvements equivalent to about 10 percent over the present state of the art in the four most important factors related to economics (L/D, air vehicle empty weight fraction, specific fuel consumption, and

TABLE 2-1 Customer Requirements, Vehicle Characteristics, and Technology Goals for Economic and Environmental Performance of Notional Aircraft

	Supersonic Business Jet	Overland Supersonic Commercial Transport	High-Speed Civil Transport	State of the Art[a]
Customer requirements				
Speed (Mach number)	1.6 to 1.8	1.8 to 2.2	2.0 to 2.4	
Range (NM)	4,000 to 5,000	4,000 to 5,000	5,000 to 6,000	
Payload (passengers)	8 to 15	100 to 200	300	
Sonic boom low enough to permit supersonic cruise over land	Yes	Yes	Yes, if possible[b]	No
Vehicle characteristics				
Payload weight fraction[c]	~0.07	0.15 to 0.20	~0.20	
Aircraft empty weight fraction[d]	~0.44	~0.40	~0.37	
Vehicle empty weight fraction[e]	~0.38	~0.34	~0.32	~0.36 (larger aircraft) to 0.38 (smaller aircraft)
Fuel weight fraction	~0.49	0.40 to 0.45	~0.43	
Takeoff gross weight (1,000 lb)	140	200 to 250	600	
Technology goals				
Economic performance				
Lift-to-drag ratio	7.5 to 8.0	9 to 10	10 to 11	~7.5 to 8.5
TSFC/M (lb/hr/lb/Mach number)[f]	~0.60	~0.52	~0.49	~0.60 (Mach 1.6) to 0.55 (Mach 2.4)
Engine thrust-to-weight ratio at sea level	5	5	6	~4 (for large engines) to 5 (for small engines)
Environmental performance[g]				
Community noise	less than Stage 3[h]	less than Stage 3	less than Stage 3	Stage 3
Sonic boom overpressure (psf)	<1 (with a shaped signature)[i]	<1 (with a shaped signature)	<1 (with a shaped signature)[b]	~2 for large aircraft, ~1 for small aircraft
NO_x emissions index at cruise (g NO_x/kg fuel)[j]	<15	<15	<15 (lower speeds), ≤5 (higher speeds)	~25[k]
Water vapor emissions index (g water/kg fuel)[l]	~1,400	~1,400	~1,400 for lower speeds, possibly 0 at higher speeds	~1,400

[a]State of the art is estimated for technologies that have matured to a TRL of 6 or higher.

[b]Only if intended for supersonic flight operations over land. Otherwise, sonic boom levels are not limiting.

[c]Weight of the payload divided by TOGW; payload is defined here as everything not necessary for controlled flight, including avionics (except the flight control system), mission equipment, and outfitting.

[d]Weight of the aircraft with no fuel or payload divided by TOGW.

[e]Weight of the aircraft with no fuel or payload *or engines* divided by TOGW.

[f]Thrust-specific fuel consumption divided by Mach number (TSFC/M) is inversely proportional to overall propulsion system efficiency. In principle, for a given propulsion system state of the art, TSFC/M varies approximately as the one-quarter power of the Mach number over the range of speed (Mach 1.6 to 2.4) of interest here. However, this has not been demonstrated in operational engines.

[g]CO_2 is an environmental constraint because it influences climate change, but CO_2 is not listed here because it is likely to be controlled by limiting total fuel consumption, not by imposing limits on emissions by individual aircraft.

[h]Current U.S. and international limits on noise for subsonic aircraft during takeoff, climb-out, and approach to landing are referred to as Stage 3 limits. Quieter Stage 4 limits are already under review.

[i]Sonic booms have a pressure wave with a very rapid rise time. Shaping of sonic booms would increase the rise time of the sonic boom pressure wave, reducing the effect of booms on people for a given overpressure limit. However, even with a shaped signature the maximum acceptable overpressure is unknown, although it seems certain to be less than 1 pound per square foot (psf).

[j]NO_x emissions are important to ozone depletion, local air quality, and climate change.

[k]State of the art for supersonic engines is about 25. Most commercial jet aircraft have an NO_x emissions index of about 7 to 15 (MCT, 2001).

[l]Water vapor emissions in the stratosphere are important to ozone depletion and climate change.

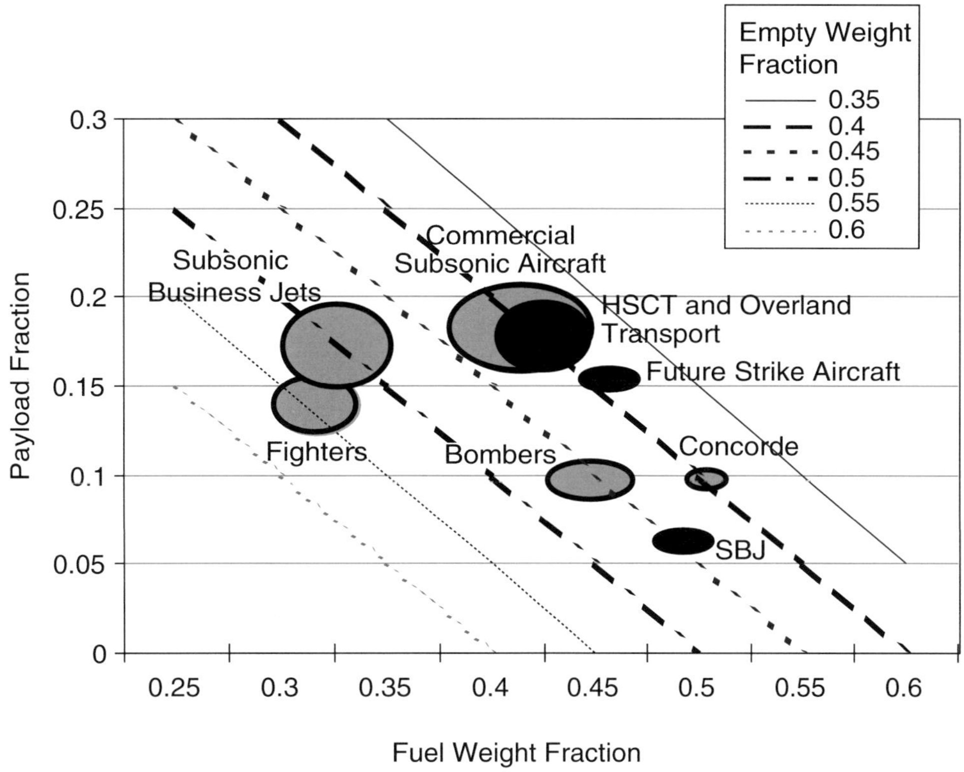

FIGURE 2-1 Weight distributions of selected aircraft.

thrust-to-weight ratio), as well as additional advances related to the environment and certification. For SBJs, most parameters are already within the state of the art. HSCTs, on the other hand, will require even more significant advances, equivalent to about a 15 percent improvement for each of the four most important economic parameters.

ENVIRONMENTAL GOALS

Real or perceived environmental requirements result in technical barriers to the development of new products. The potential uncertainty associated with environmental effects and regulations pose nontechnical barriers as well, because industry is unlikely to commit to product development without certain knowledge of the environmental standards that must be satisfied. Some of the relevant environmental regulations are summarized in Table 2-2. In general, existing regulations were not developed with supersonic aircraft in mind and would have to be updated before a new commercial supersonic aircraft could be certificated. All of these issues will have to be coordinated domestically with the appropriate federal agencies and internationally with the International Civil Aviation Organization (ICAO).

Technology goals associated with environmental performance of commercial supersonic aircraft are shown in Table 2-1. As discussed below, the major environmental concerns

TABLE 2-2 Environmental Regulations Relevant to Commercial Supersonic Aircraft

Environmental Issue	Current or Expected Method of Control	International Regulations and Authorities	U.S. Regulation and Authorities
Community noise	Aircraft certification standards, operating restrictions	ICAO (Annex 16, Vol. I)	14 CFR Part 36 and 14 CFR Part 91
Sonic boom	Operating restrictions	ICAO (Resolution A33-7)	14 CFR Part 91
Climate change	Aircraft certification standards, market-based measures (emissions trading or charges)	United Nations Framework Convention on Climate Change and ICAO (under Kyoto Protocol, if ratified)	14 CFR Part 34
Ozone depletion	Operating restrictions	Montreal Protocol	Section 615 of the Clean Air Act
Local air quality	Aircraft certification standards	ICAO (Annex 16, Vol. II)	14 CFR Part 34 and 40 CFR Part 87

are community noise around airports, sonic boom (which prevents supersonic flight over land), climate change, depletion of atmospheric ozone, and local air quality. Reducing sonic boom enough to meet the needs of SBJs or medium-size overland transports could be very difficult. Continuation of ongoing research is likely to reduce emissions enough to meet goals both for local air quality and for emissions during cruise for aircraft with cruise speeds less than approximately Mach 2. Cruise emissions goals for aircraft with higher cruise speeds will be considerably more difficult to meet, and noise suppression will need attention at all Mach numbers.

Community Noise

Commercial aircraft must meet community noise standards for takeoff, climb-out (after takeoff), and approach to landing. For subsonic aircraft, requirements are defined by so-called Stage 3 standards. ICAO standards for supersonic aircraft are not in place, but ICAO Annex 16 Volume I indicates that "noise levels . . . applicable to subsonic jet aeroplanes may be used as guidelines." Except for existing Concorde aircraft, federal regulations prohibit the operation of commercial supersonic aircraft unless they comply with Stage 2 noise limits (14 CFR Part 91). Federal regulations allow existing Concorde aircraft to operate as long as (1) the noise levels are shown to have been reduced "to the lowest levels that are economically reasonable, technologically practicable, and appropriate for the Concorde type design," and (2) the aircraft does not take off or land between the hours of 10 p.m. and 7 a.m. (14 CFR Part 36).

Environmental standards have historically been governed by what can be accomplished within the economic constraints of the industries involved, and more stringent standards are already under development by ICAO's Civil Aviation Environmental Protection (CAEP) Committee: Stage 4 limits are likely to reduce noise limits during takeoff, climb, and approach to landing by about 3 dB at each point. Presumably, these new ICAO limits would be used as guidelines for revising community noise limits in the United States and other countries, but details, including how these standards would be applied to supersonic aircraft—would have to be coordinated with regulatory agencies, such as the FAA, in each country.

Sonic Boom

ICAO policy requires aircraft operators to ensure that "no unacceptable situation for the public is created by sonic boom from supersonic aircraft in commercial service" (ICAO Resolution A33-7, 1998). This seems to leave room for determining an "acceptable" level for a sonic boom. U.S. Federal Aviation Regulations (FARs), however, prohibit commercial flight at greater than Mach 1 over the United States and require that commercial supersonic aircraft, such as the Concorde, that fly to and from the United States impose flight restrictions to ensure that they do not "cause a sonic boom to reach the surface within the United States" (14 CFR Part 91).[1] These regulations will have to be changed to allow supersonic flight over land. Except for supersonic speeds close to Mach 1 (i.e., Mach numbers between 1 and about 1.15), sonic boom reduction technology is unlikely to eliminate sonic booms in the foreseeable future. As discussed in Chapter 3, a more reasonable aim for sonic boom reduction is to produce "shaped" sonic signatures with an overpressure limit somewhere between 0 and 1.0 lb/ft^2 throughout the total sonic boom ground footprint (not just below the aircraft). Low-boom, shaped signatures are intended to reduce annoyance and minimize the possibility of damaging structures by reducing startle, rattle, and building vibrations. Additional research is needed, however, to ascertain the limits on sonic boom parameters, such as overpressure and rise time, that would be acceptable to the general public. Research must also validate the ability to design aircraft that (1) create shaped booms within acceptable limits and (2) still meet other economic and environmental goals. These advances are necessary for both the SBJ and the medium-size overland transport, since both require flight over land and would be beneficial for the HSCT.

Climate Change

Under the Kyoto Protocol, ICAO was given the responsibility for control of aircraft emissions that affect climate change. Accordingly, CAEP is examining methods to control aircraft emissions that can affect climate. The primary focus now is carbon dioxide (CO_2). The Kyoto Protocol did not classify water or oxides of nitrogen (NO_x) as greenhouse gases, but their treatment is being discussed in ICAO. Because of water vapor emissions, atmospheric models generally predict that an aircraft cruising in the stratosphere can have climate effects several times as severe as a similar aircraft flying at lower altitudes (in the troposphere), where water vapor emissions are benign.

Emissions of CO_2 are directly related to fuel consumption, and the methods of control being considered by CAEP are all market-based options, such as a system of CO_2 trading (similar to current SO_2 trading programs in the United States) or fees based on fuel consumption. If such programs are implemented in a way that increases fuel costs, they would increase economic barriers for supersonic aircraft, which consume more fuel per passenger seat mile than sub-

---

[1]The FAA is authorized to allow supersonic flight over land for testing—if over-water flights are not practical (14 CFR Part 91). Also, military aircraft may fly at supersonic speeds in authorized training areas, but environmental impact statements may be required to establish or expand training areas if supersonic flight will be allowed at altitudes less than 30,000 ft over land (or within 15 miles of land) or at altitudes less than 10,000 ft over water (32 CFR Part 989).

sonic aircraft.[2] In addition, uncertainties about how CO_2, water vapor, and NO_x may be regulated in the future add to uncertainties about the economic viability of commercial supersonic aircraft.

ICAO is currently developing a methodology to characterize aircraft emissions during the climb and cruise phases of flight, and the initial focus is on NO_x emissions. The methodology will consider emissions and aircraft productivity, probably in a format such as grams of pollutant per passenger-mile or ton-mile of payload. Details of the methodology and a typical range of values for subsonic aircraft are expected to be discussed at the sixth CAEP meeting, which will probably occur in 2004, and a standard could be established in the 2010 to 2015 time frame. Supersonic aircraft developed during and after that time may be required to meet the same standards as subsonic aircraft. That could be difficult, because the increased fuel consumption and higher engine operating temperatures that occur during supersonic cruise tend to increase NO_x emissions.

Ozone Depletion

The stratospheric ozone layer absorbs ultraviolet radiation that can cause skin cancer. Many compounds that deplete the ozone layer, such as chlorofluorocarbons, are controlled under the Montreal Protocol. NO_x emissions by aircraft can impact atmospheric ozone in either positive or negative ways, depending on the altitude at which they are emitted. There is considerable uncertainty in the estimates of the actual impact, but it is generally agreed that emissions at altitudes above 50,000 to 55,000 ft will degrade the ozone layer to some degree. Concern about ozone destruction by NO_x emitted from supersonic aircraft flying in the stratosphere dates back to the U.S. SST program in the early 1970s and was acknowledged by the HSR Program, which established an NO_x emissions index goal of 5 (i.e., 5 g of NO_x produced for each kilogram of fuel burned during cruise). This still appears to be an acceptable upper limit for commercial supersonic aircraft operating in the stratosphere. For aircraft at lower supersonic speeds (i.e., less than approximately Mach 2), NO_x emissions would be less of a problem because of their lower cruising altitudes, and a higher emissions index might be acceptable.

Recent analyses (e.g., Kawa et al., 1999) indicate that water emissions from a fleet of large commercial supersonic aircraft operating in the stratosphere would significantly deplete ozone, even if NO_x could be reduced to very low levels. Although the analytical results have considerable uncertainty, effects would be substantially smaller for a smaller fleet of smaller aircraft and could be eliminated by flying at a lower altitude consistent with a lower cruise speed.

Local Air Quality

ICAO standards for both subsonic and supersonic aircraft cover emissions of NO_x, unburned hydrocarbons (HC), carbon monoxide (CO), and visible smoke. Standards for supersonic aircraft were based on the Concorde emissions levels and therefore apply to an engine that uses an afterburner during takeoff. Engines with afterburners generally produce more CO and less NO_x than typical subsonic jet engines. Presumably, the next generation of supersonic transports will not use afterburners, so emissions should be more like those of subsonic aircraft. Since the current supersonic standard was set, there have been substantial reductions in emissions of all of these species, and NO_x standards for subsonic aircraft has been reduced twice. If a new supersonic aircraft is introduced into service, ICAO standards would very likely be reexamined in light of these improvements in subsonic engine technology. To minimize opposition on environmental grounds, new supersonic aircraft may be required to meet the same standards as subsonic aircraft.

TECHNOLOGY CHALLENGES

The key technology challenges that derive from the customer requirements, vehicle characteristics, and technology goals quantified in Table 2-1 are related either to economics, the environment, or certification:

- Environment
  - —benign effect on climate and atmospheric ozone
  - —low landing and takeoff noise
  - —low sonic boom
- Economics—range, payload, fuel burn, etc.
  - —low weight and low empty weight fraction
  - —improved aerodynamic performance
  - —highly integrated airframe/propulsion systems
  - —low thrust-specific fuel consumption (TSFC)
  - —long life
- Certification for commercial operations
  - —acceptable handling and ride qualities
  - —passenger and crew safety at high altitudes
  - —reliability of advanced technologies, including synthetic vision
  - —technical justification for revising regulations to allow supersonic operations over land

These challenges are discussed in more detail in Chapters 3 and 4. All of the challenges must be overcome to enable viable commercial supersonic aircraft. Significant advances in the traditional aeronautics engineering disciplines, such as structures, propulsion, and aerodynamics, are still required

[2] Passenger seat mile is a unit of transportation capacity, whereas passenger mile is a measure of service provided. An aircraft with 100 seats scheduled to fly a 200-mile route would have a capacity of 2,000 passenger seat miles. If 50 passengers make the trip, only half of the capacity is used (1,000 passenger miles).

to create a situation in which (1) the economic factors will support a profitable business case and (2) certification issues for new designs and systems can be resolved. Of the various vehicle types examined by the committee, the technology advances necessary to attain the required economic and environmental goals will be easiest to achieve for an SBJ and most difficult for an HSCT. The economic goals for the SBJ are easiest to achieve because it can tolerate a very low payload fraction, a shorter range capability, and a modest cruise speed. Economically viable sonic boom reduction is also likely to be easier to accomplish for an SBJ than for a medium-size overland transport. The economic goals for the HSCT will be the most difficult to achieve because it needs the highest payload fraction and the longest range. Furthermore, if a cruise speed greater than approximately Mach 2 is selected, the environmental goals would be much harder to achieve.

The above discussion has been limited to aircraft powered by gas-turbine engines using conventional hydrocarbon fuels. The potential use of other power plants and/or other fuels would change the tradeoffs among the various technology parameters, but it would be unlikely to reduce the gap between what is needed and what is currently available. For example, the use of liquid hydrogen as a fuel has been examined in the past and is a prospect for the future. Because hydrogen offers high energy per unit mass of fuel, it reduces TSFC to about 40 percent of the consumption for a conventional hydrocarbon fuel. However, hydrogen's low density requires large fuel storage volumes, which would increase the air vehicle empty weight fraction, decrease the L/D, and increase sonic boom. Thus, in the terminology used here, the technology requirements would change, but the challenges would not diminish. Similar observations can be made about the potential use of fuel cells as aircraft power plants. In principle, fuel cells offer perhaps a 20 percent reduction in TSFC compared with turbofan engines, but at the expense of a considerably heavier power plant. Accordingly, the needed values of TSFC and thrust-to-weight ratio will change (the former will be lower and the latter higher than those shown in Table 2-1), but the challenges will not diminish.

Certification and regulatory issues are especially important in two respects: (1) to enable the use of new systems, such as synthetic vision in the cockpit, that relate to passenger and public safety and (2) to allow overland supersonic flight if an aircraft with a low, but nonzero, sonic boom can be developed.

REFERENCES

Kawa, S., J. Anderson, S. Baughcum, C. Brock, W. Brune, R. Cohen, D. Kinnison, P. Newman, J. Rodriguez, R. Stolarski, D. Waugh, and S. Wolfsy. 1999. Assessment of the Effects of High-Speed Aircraft in the Stratosphere. Greenbelt, Md.: Goddard Space Flight Center.

MCT (Ministério da Ciência e Tecnologia)(Brazil) 2001. Inventory of Emissions per Type of Aircraft in 1995. Available online at <http://www.mct.gov.br/clima/ingles/comunic_old/iac01.htm>. Accessed on July 10, 2001.

NRC (National Research Council), Aeronautics and Space Engineering Board. 1997. U.S. Supersonic Commercial Aircraft: Assessing NASA's High Speed Research Program. Available online at <http://www.nap.edu/catalog/5848.html>. Accessed on September 13, 2001.

3

New Opportunities for Research on Critical Supersonic Technologies

The development of economically viable and environmentally acceptable commercial supersonic aircraft will require continuing advances in many disciplines. This chapter deals with technology areas where important breakthroughs are possible only if current research efforts are augmented by new, focused research (or significant expansions of existing research). The committee identified five areas of critical importance: approaches to sonic boom reduction, new aerodynamic concepts to improve efficiency and reduce environmental impacts, methods for dealing with highly coupled aircraft dynamics, strategies for the design of complex systems, and the continued development of variable cycle engines.

CONFIGURATIONS FOR REDUCED SONIC BOOM

Background

Shock waves, and thus sonic booms, are fundamental to supersonic flight and can be minimized, but not eliminated, on aircraft having lift. Sonic booms can be startling, cause annoyance, and can even result in structural damage. Current U.S. regulations prohibit commercial aircraft from producing sonic booms that can be detected on the ground. If an approach to vehicle aerodynamic design can be found that would result in sonic booms with intensities low enough to gain public acceptance—and if regulations are changed to allow low intensity sonic booms—the economics of supersonic flight would change dramatically.

Studies by NASA's HSR Program identified three key requirements for overland supersonic flight: (1) establishing the criteria for an acceptable "shaped" sonic boom signature, (2) designing a viable aircraft to produce that shaped signature, and (3) quantifying the influence of the atmosphere on such signatures.[1] However, no revolutionary aerodynamic approaches to sonic boom elimination seem imminent, and the feasibility of accomplishing step 2 has yet to be established.

Nearly 50 years of flight data and experience with sonic booms exist, covering some 20 different supersonic aircraft, including the Concorde and the space shuttle, with over 1,500 flights having produced some 15,000 measured signatures. All of these sonic boom signatures are sawtoothed (N waves), since all the aircraft were so-called N-wave designs—that is, sonic boom minimization was not considered in their basic design. Current signature prediction codes, validated by numerous wind-tunnel model tests, in-flight flow field probes, and ground measurements, work quite well for these N-wave aircraft.[2]

Fairly substantial efforts by the government, industry, and universities in the mid-1960s and 1970s and, later, during the HSR Program explored boom minimization techniques that decrease overpressure and shape the signature by means of aircraft tailoring and airstream alteration. Vehicle configurations designed for boomless flight were also investigated. However, no flight data have ever been collected on vehicles designed to produce low-boom, shaped signatures. Wind tunnel models can be designed to produce booms with shaped signatures near the aircraft, but it has not been demonstrated, by analysis or experiment, that a shaped signature will persist to the ground during flight in a real atmosphere. Additional research is needed to establish a credible scientific foundation for designing supersonic aircraft with low sonic booms and to develop improved analytical tools.

It is well known that atmospheric turbulence in the lower layers of the atmosphere can produce large changes in the overpressure or intensity of N-wave sonic boom signatures.

[1]"Signature" refers to a plot of the change in air pressure versus time at a fixed point as a sonic boom passes.

[2]An N-wave signature has two rapid increases in pressure, one at the beginning and one at the end of the sonic boom. This produces a loud and particularly startling double-boom for people in the affected region. Shaping the signature so that it has a slower pressure rise would produce a less intense boom.

Although considerable progress has been made toward establishing a prediction capability to describe the statistical variations of N-wave signatures due to the atmosphere, a suitable prediction code does not exist. No analytical or experimental database exists for quantifying atmospheric effects on low-boom, shaped signatures. Model experiments using a large ballistic range, small projectiles, and a rectangular jet flow nozzle to generate scaled turbulence have provided valuable information. Similar tests using shaped projectiles designed to produce shaped booms could generate most of the required database to determine the influence of the lower layer of atmosphere on shaped boom signatures. An important milestone is scheduled for September 2002: Northrop Grumman, as part of DARPA's QSP Program, plans to flight test an F5E aircraft modified to produce a shaped signature below the aircraft when flown at Mach 1.4 and 30,000 ft. Experiments using this vehicle will help determine the extent to which analytical tools used to predict the propagation of N-wave signatures can be used for shaped signatures.

Most of the data on the response of people and buildings to sonic booms are for N-wave signatures and boom overpressures greater than 1.0 lb/ft^2. There is no database on the response of people and buildings to shaped signatures of less than 1.0 lb/ft^2. Human response data are also needed for nighttime booms, both to guide the development of new technology and to support necessary changes in federal regulations that prohibit sonic booms over land (see Chapter 2). A significant finding from past sonic boom studies is that startle, rattle, and building vibrations (which can cause damage) are key elements in determining the response of people to sonic booms. Shaped signatures of less than 1.0 lb/ft^2 will produce less startle, rattle, and building vibrations.

Potential Developments

Key requirements for designing a viable supersonic aircraft having a low-boom, shaped signature are low weight, high L/D, long length, and innovative propulsion integration. Focused research on vehicle configurations to produce shaped waves, combined with incremental improvements in existing technologies, may lead to vehicle configurations that produce shaped signatures with a maximum amplitude of less than 1.0 lb/ft^2 throughout the total sonic boom ground footprint (not just below the aircraft). Studying the human response to shaped waves will also be necessary, both to assist vehicle design research and to validate new regulatory standards, which may require reducing sonic boom amplitude significantly below 1.0 lb/ft.

Research Opportunities

Research on methods for sonic boom prediction and techniques for designing aircraft with specified or constrained signatures has been ongoing for decades, but new approaches that permit higher fidelity modeling and more efficient design are required. Research should focus on the continuing development of (1) multidisciplinary design tools and (2) one or more flight technology testbeds to characterize the booms produced by shaped vehicles and measure the persistence of shaped signatures in a real atmosphere over large distances. Laboratory and wind tunnel testing on vehicle shaping and atmospheric effects should also be conducted, including some testing on airstream alteration (e.g., heat/energy addition and dynamic flow modifications) for a variety of vehicle configurations. Finally, community studies are essential to determine how shaped sonic boom signatures with overpressures less than 1.0 lb/ft^2 effect buildings and people, including the propensity of people to be roused from sleep by booms at night.

ADVANCED AERODYNAMIC CONCEPTS AND CONFIGURATIONS

Background

The character of flow over a vehicle in supersonic flight is dramatically different from that of a subsonic aircraft. This is one of the fundamental reasons that an economically viable, environmentally acceptable supersonic aircraft has not been achieved after more than a half century of work in aeronautical design. Breakthrough technologies that could address the root causes of the difference will therefore be associated with the vehicle aerodynamics. This section deals with aerodynamic challenges and opportunities for new research that may lead to dramatic improvements in vehicle performance and efficiency.

Aerodynamic cruise efficiency is extremely important because it directly and indirectly impacts most of the challenges faced by the development of a viable commercial supersonic aircraft. This is demonstrated clearly in the case of the Concorde, which has a cruise L/D of 7.5 (modern subsonic transports have a cruise L/D of about 18 to 20). The low L/D of the Concorde increases fuel consumption, limits its range, increases the design takeoff weight, and requires a larger propulsion system to provide the higher thrust required at takeoff, which in turn makes it more difficult to meet community noise standards. With engines that are as efficient as those of modern subsonic aircraft, the Concorde would still have to carry about three times the weight and generate more than three times the takeoff thrust of a 737-600, which can carry the same number of passengers the same distance.

The low supersonic efficiency of the Concorde, which has a cruise speed of Mach 2, is not a result of poor design but a fundamental consequence of supersonic aerodynamics. For a supersonic aircraft, the maximum achievable L/D depends on several aircraft characteristics and drops as Mach

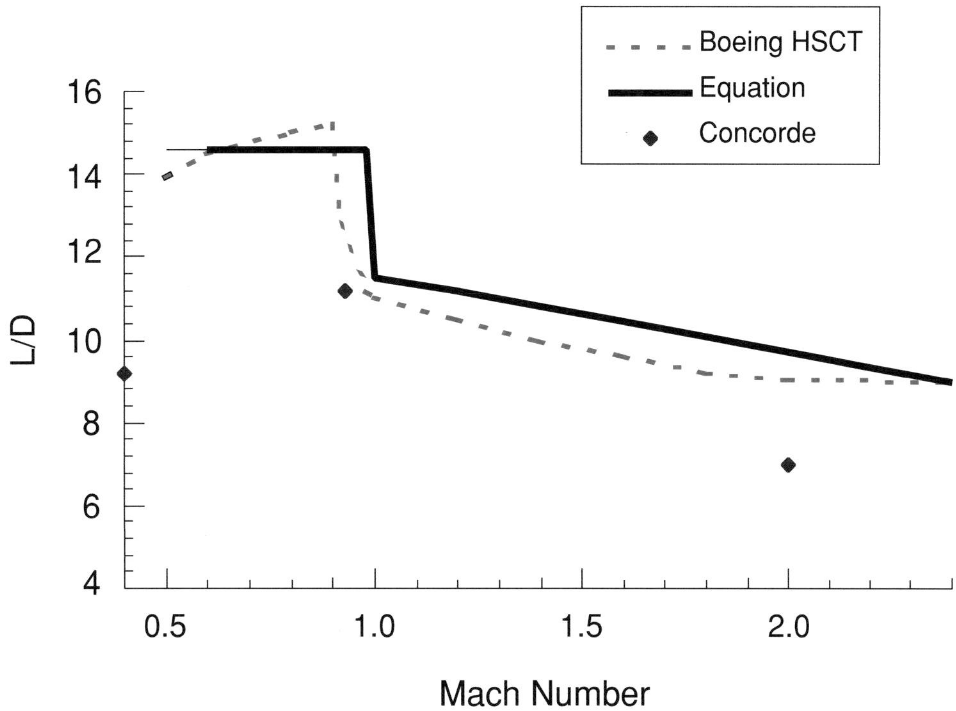

FIGURE 3-1 Typical variation in L/D with Mach number. SOURCE: Boeing (1989).

number is increased. Figure 3-1 shows the typical variation in L/D with Mach number.[3]

The requirement for low cruise drag generally leads to supersonic vehicles that are much longer than subsonic aircraft with a comparable payload capacity. This, in turn, leads to penalties in the form of degraded vehicle performance at low speed and increased structural weight. As the cruise Mach number is increased, the optimal slenderness of the vehicle increases as well, creating even further disparity between the best cruise design and the best low-speed design. Another critical design conflict arises because the optimum engine bypass ratio decreases as Mach increases. Design compromises result in degraded subsonic performance and lower usable maximum lift during low-speed flight. For example, the Concorde achieves an L/D of only about 5 to 9 in low-speed flight (Rech and Leyman, 1980). As a result, almost 40 percent of the fuel carried at takeoff is devoted to low-speed flight and reserve fuel.

Offsetting the difficulties associated with higher Mach numbers is the increased utilization (distance flown per year) that higher speeds make possible. Also, the range achievable in long-range cruise at constant speed is proportional to

$$\frac{V\,(\mathrm{L/D})}{C}\log\left(\frac{W_1}{W_2}\right)$$

where V is the cruise speed, C is the specific fuel consumption, and W_1/W_2 is the ratio of the initial weight (at takeoff) to the final weight (at landing). Compared with more aerodynamically efficient subsonic aircraft, the aerodynamic penalty (in terms of lower L/D) for flying at Mach 2.0 to 2.4 is about 10 percent greater than the penalty at Mach 1.6 to

[3]The minimum drag of a supersonic aircraft may be expressed as follows (Jones and Cohen, 1960):

$$D = qSC_{D_0} + \frac{W^2}{q\pi b^2} + \frac{128qVol^2}{\pi l^4} + \frac{(M^2-1)W^2}{q\pi l^2} \quad (1)$$

where b is the wingspan, C_{D_0} is the parasite drag at zero lift (i.e., skin-friction drag and all other drag except induced drag), l is the effective length of the vehicle, M is the Mach number, q is the dynamic pressure, S is the reference wing area, Vol is the total volume, and W is the total weight of the aircraft.

If the altitude is chosen to be optimal at the given Mach number, the L/D can be written as follows:

$$\mathrm{L/D}_{\max} = \left[\left(\frac{4}{\pi AR} + \frac{2(M^2-1)}{\pi AR_1}\right)\left(C_{D_0} + C_{D_{0w}}\right)\right]^{-1/2} \quad (2)$$

where AR, the aspect ratio, is given by b^2/S; AR_l, the length aspect ratio, is l^2/S; and $C_{D_{0_w}}$, the zero lift wave drag, is a function of aircraft length, volume, and wing area. Equation 2 is plotted in Figure 3-1, assuming that C_{D_0} and $C_{D_{0_w}}$ have values of 0.01 and 0.006, respectively, and that AR and AR_l have values of 2.7 and 10, respectively. (The wave drag terms, $C_{D_{0_w}}$, and the term containing AR_l are set to 0 below Mach 1.)

1.8 (see Figure 3-1). As the cruise speed of a supersonic aircraft increases, specific fuel consumption also increases. However, in principle, the parameter *V*/*C* actually increases with speed by a factor of about $V^{1/4}$ over the range of interest here. An increase of this magnitude would largely offset the drop in L/D that occurs at higher speeds. However, this increase has not been demonstrated in operational engines. Furthermore, parametric design studies of supersonic airplanes show that designers have much less freedom to make necessary compromises than with subsonic aircraft. Thus, a net reduction of a few percent in [$V/C \times L/D$] may have a serious cascading effect on the ability to meet other design goals at higher speeds. Regardless of the cruise speed of interest, increases in L/D are vital to enabling an economically viable supersonic aircraft, and meeting L/D goals (see Table 2-1) should remain the focus of aerodynamic research for commercial supersonic aircraft.

Most work to date has focused on the higher speeds to increase the benefit in terms of utilization. Based on more recent estimates of environmental constraints, however, a range of 1.4 to 2.0 seems more reasonable and would open the possibility for new benefits from advances in vehicle configuration and aerodynamics. As described in the preceding section, improved vehicle configuration designs are also likely to be a key part of reducing sonic boom to levels that might permit overland supersonic flight.

State of the Art

Research in supersonic aircraft aerodynamics, which has been ongoing for almost 50 years, has been marked by intermittent efforts—first in the late 1960s and early 1970s when the Concorde was developed and a U.S. SST was worked on, then again in the 1990s as part of NASA's HSR Program and concurrent work by U.S. industry to develop designs for an HSCT. Currently attention is being given to smaller supersonic aircraft by the DARPA QSP Program, and industry (e.g., Gulfstream and Dassault) is exploring the development of an SBJ (George, 2000).

One view of the progress that has been achieved is shown in Figure 3-1, which compares the L/D of a recent Boeing HSCT design with that of the Concorde. Although the estimates for the HSCT are somewhat optimistic, the improvements in cruise performance and particularly in subsonic performance are significant. Although a change in cruise L/D from 7.5 to 9 represents only a 20 percent improvement in drag at a given weight, supersonic aircraft are more sensitive to such changes than are conventional subsonic aircraft. For a design range of 5,000 NM and with an assumed engine efficiency of 45 percent and an empty weight fraction of 0.25, a 20 percent increase in L/D corresponds to a 40 percent reduction in the required takeoff weight for a 100-passenger aircraft. Several promising concepts, while immature, may become key features of a successful future commercial supersonic aircraft.

Related Promising Technologies

Supersonic aerodynamics could be revolutionized by successful technologies in any of four areas: supersonic laminar flow, other methods for modifying the flow field around the aircraft, unconventional vehicle configurations, and detailed, computational systems for high-fidelity analysis.

Supersonic laminar flow has long been recognized as a potential breakthrough that might reduce skin friction drag by as much as 90 percent. But achieving extensive laminar flow has been an elusive goal. Substantial efforts are being made to achieve laminar flow for subsonic aircraft, but results have not been particularly encouraging. Research on suppressing the transition from laminar to turbulent flow using active flow control (via suction, blowing, or time-dependent boundary-layer manipulation) continues in many laboratories, but the prospect of developing an economically viable system of this sort remains remote. Perhaps more intriguing is the possibility that laminar flow may be more easily maintained at supersonic speeds than at lower speeds. A few related approaches involve the careful design of wing surfaces to achieve favorable streamwise pressure gradients and minimize cross-flow transition. These approaches range from those described by Tracy et al. (1995), in which wings with low leading-edge sweep and favorable chordwise pressure gradients are integrated into the aircraft concept, to recent work at the National Aerospace Laboratory of Japan, which emphasizes more highly swept wings with low upper-surface cross-flow achieved with rather flat streamwise pressures (Yoshida et al., 2000). The latter concept, while more sensitive to disturbances and aimed at achieving laminar flow over 25 percent of the wing surface, permits the use of substantial wing sweep. The more mildly swept natural laminar flow concept has demonstrated much larger extents of laminar flow in recent flight tests but may incur structural penalties because of the need for very thin wings. In addition, the short lifting length of the reference concept is difficult to reconcile with the requirement for shaped sonic boom signatures. Research on each of these concepts is in its infancy but, if successful, may have a dramatic effect on achievable supersonic aircraft performance. It appears feasible to extend these ideas with additional measures for cross-flow suppression. Active cooling or passive techniques for suppressing the initial cross-flow instability (White and Saric, 2000) may permit additional sweep on the supersonic leading-edge natural laminar flow concept or more extensive laminarity for the subsonic concept.

The second area of general interest for dramatic improvements in supersonic aerodynamics involves much more speculative approaches to the modification of the flow field. Active flow control, virtual shaping, and energy addition in various forms have been proposed for many years as a possible means for reducing wave drag or sonic boom amplitude. The committee does not believe that any of these approaches promise near-term breakthroughs in supersonic performance or boom reduction. Indeed, some of them appear to have no reasonable physical basis, while others are either too complex to evaluate at this time or feature concepts whose practical implementation is hard to imagine (see, for example, Rethorst and Kantner, 1996; Rising and Vadyak, 1999; and Soloviev et al., 1999). Future research programs should consider investigating such concepts, with the understanding that success, while unlikely, would be important.

A third alternative is to investigate unconventional designs that attempt to address some of the fundamental problems encountered with supersonic aerodynamics. These range from nonplanar and multiple-surface configurations to asymmetric, oblique wings. Several of these ideas are based on fundamentally sound aerodynamics, but integrating them into practical aircraft designs has been difficult. In some cases, this is due to a basic limitation of the concept. For example, the oblique all-wing concept that accommodates passengers inside the wing structure appears to offer spectacular aerodynamic performance and great potential for reducing sonic boom, but it is difficult to configure as a passenger aircraft unless it is scaled up to accommodate 500 passengers (Jones, 1991; Seebass, 1994). In other cases, the complexity of the configuration may limit the applicability of simple analyses, and the associated risk and large amount of work required to develop appropriate analysis methodologies cannot be accommodated within the time and resource constraints of ongoing supersonic research programs. Immature vehicle configurations also have a hard time competing against vehicle configurations that have long histories of wind tunnel testing and computational design analyses. The solution to this dilemma may lie in the development of analytical methods that permit higher-fidelity analysis of new concepts early in the design cycle.

High-fidelity analysis of unconventional vehicle configurations is just now becoming feasible and represents a true opportunity for breakthrough technology. Advances in computational algorithms for aerodynamic analysis and shape optimization, together with a revolution in computer hardware capabilities, now make it possible to consider a much wider range of design possibilities at a level of detail formerly restricted to a single baseline design. More mature flow solvers, improved representations of boundary layer turbulence, and methods for efficient calculation of flow field sensitivities to design changes make the evaluation of alternative design concepts feasible. Coupled with advances in techniques for multidisciplinary optimization, such capabilities hold out the promise that unconventional concepts can be transformed into practical breakthrough technologies. As an example, consider the concept of natural laminar flow. Until recently, the ability to predict transition of a three-dimensional boundary layer and use this prediction to design a wing with extensive laminar flow was a remote possibility. Indeed, tests on an F-104 in the late 1950s showed that limited laminar flow could be achieved, but tools were not available to analyze the results, let alone use them to design a wing. More recently, a specified wing design was analyzed to assess its potential for extensive natural laminar flow (Agrawal and Powell, 1991). The conclusion was that despite the small sweep, little laminar flow would occur. Current computations including nonlinear computational fluid dynamics, three-dimensional boundary layer analysis, and stability calculations have made it possible to successfully optimize a wing for extensive laminar flow. Combining this capability with structural analysis and more comprehensive aircraft performance calculations would greatly advance the prospects for using natural laminar flow to significantly improve the performance of a commercial supersonic aircraft.

Basic and Applied Aeronautics Research

Future research programs that would support the development of the technologies described above should include the following:

- techniques to predict and control the transition from laminar to turbulent flow, including supersonic natural laminar flow, passive control techniques, cooling, periodic roughness, and active control methods
- improved, design-oriented computational fluid dynamics for improved multipoint performance and sonic boom reduction, including adaptive, unstructured methods that achieve the efficiency of current multigrid structured methods and low dissipation methods that can be used in combination with boom propagation codes
- multidisciplinary design methods compatible with high-fidelity modeling, as well as optimization, decomposition, and methods that exploit parallel computing architectures
- flight R&T demonstrations to help investigate technologies of interest that remain difficult to validate computationally, or even in existing wind tunnel facilities (NRC, 1994)

Flight experiments will play an important role in the development of new concepts, and methods for more efficient flight tests, including the development of sensors for flow diagnostics, will be especially important. New supersonic wind tunnel capability may also be needed.

VEHICLE DYNAMICS AND CONTROL

Problem Description

To achieve the required aerodynamic performance, next-generation supersonic transport aircraft will likely exhibit greater aerodynamic instabilities than any other existing or planned transport aircraft. These instabilities must be stabilized using high-authority, flight-critical feedback control systems. Furthermore, owing to their size and shape, large commercial supersonic aircraft will also exhibit unusually low structural-vibration modal frequencies. For example, the 1- to 1.5-Hz first-fuselage modes predicted for the baseline HSCT designs considered by the HSR Program would be significantly lower than those for any manned aircraft, commercial or military (NRC, 1997).

Either aerodynamic instability or structural flexibility considered separately would present significant technical challenges. Taken together, they create severe frequency coalescence (between the rigid-body and structural modes) that must be carefully examined before developing the flight- and structural-mode control systems. The combination of aerodynamic instability and structural flexibility also creates two forms of multidisciplinary feedback phenomena involving the elastic airframe, flight-control systems, and either the pilot or the aircraft's propulsion systems. Hence, both phenomena are encompassed by the acronym APSE, which can refer to either aircraft-pilot servo-elastic or aero-propulsive servo-elastic phenomena. Although both APSE phenomena fundamentally depend on structural deformations, APSE is not an aeroelastic problem per se and it cannot be solved by methods used to counter classical aeroelastic flutter.

The aero-propulsive servo-elastic phenomenon was discussed in some detail in a previous NRC report that reviewed NASA's HSR program (NRC, 1997). That report concluded that this phenomenon was "completely outside industry's experience base," and the HSR program had not found a solution to the problems created by the phenomenon.

Regarding the aircraft-pilot servo-elastic phenomenon, the excitation of the elastic modes of aircraft with structural modal frequencies below 2 Hz (which is a natural consequence of turbulence and pilot control inputs) will be orders of magnitudes greater than that encountered in other transport aircraft. If left unmitigated, these modal excitations would create unacceptable handling quality and ride quality. In particular, pilot excitation of low-frequency structural modes, coupled with the pilot's biodynamic response to these excitations, has been recently demonstrated to lead to aircraft-pilot coupling instabilities (see Box 3-1) (Raney et al., 2001).

Configuration Design Implications

Both forms of APSE phenomena are exacerbated by vehicle configurations with long and slender fuselages and thin or highly swept wings, by lightweight (hence low stiffness) structural design, and by increased aerodynamic instability, all of which are key factors in achieving high-performance, low-boom commercial supersonic aircraft. Thus, even though APSE effects are likely to be a major factor in defining the vehicle configuration for next-generation commercial supersonic aircraft, their importance in this regard is not generally recognized and they are rarely incorporated into the early stages of aircraft design. Consequently, research into APSE phenomena and their causes, along with new analysis and synthesis tools, is required. These tools include new active-control concepts, new control-system synthesis techniques, and aeroelastic modeling approaches that may be used in the vehicle configuration-design phase and integrated into multidisciplinary optimization techniques.

Additional Research Required

Low-frequency structural vibration modes will require active structural mode control systems that are highly integrated with the primary flight-control systems. Success in this effort will be particularly beneficial because solving the control problems associated with low-frequency structural vibration modes is one key to resolving both forms of APSE phenomena. New techniques must be developed and validated for designing affordable, certifiable, highly integrated, high-authority flight- and structural-mode control systems. Research in handling qualities is necessary to develop design criteria for aircraft control systems. Additionally, novel sensors and actuation devices, along with novel distributed control approaches, must be considered.[4,5]

New structural design approaches are also required, and new tools and technology must be developed and validated for designing affordable, advanced structural systems that can be certificated for use on commercial aircraft. Options include (1) multidisciplinary configuration optimization techniques that capture both types of APSE phenomena, (2) active or smart structures, and (3) viscoelectric or electrorheological materials.[4,5]

Finally, current practice in the industry creates functionally separate flight controls engineering organizations and structural dynamics engineering organizations. This is not an appropriate organizational structure for handling the issues associated with these APSE phenomena. As previously recommended by the NRC in an assessment of the HSR Program, interdisciplinary teams should be formed to fully address relevant aspects of the APSE problem, and the organizational distance between groups responsible for (1) guidance and control systems and (2) structural-mode control laws should be reduced or eliminated (NRC, 1997).

Experimental programs should be initiated as a first step in establishing better handling- and ride-quality requirements for highly unstable and highly flexible aircraft and associated flight control and structural-mode control systems. Real-time, manned simulations of the dynamics of these vehicles, however, would severely tax both fixed-base and in-flight simulation facilities, limiting the ability to establish

[4]Anna-Maria McGowan, NASA Langley Research Center, personal communication with David Schmidt, 2001.

[5]Terrance Weisshaar, Purdue University, personal communication with David Schmidt, 2001.

BOX 3-1 Demonstration of Aircraft-Pilot Coupling Instability in a Ground-Based Simulation

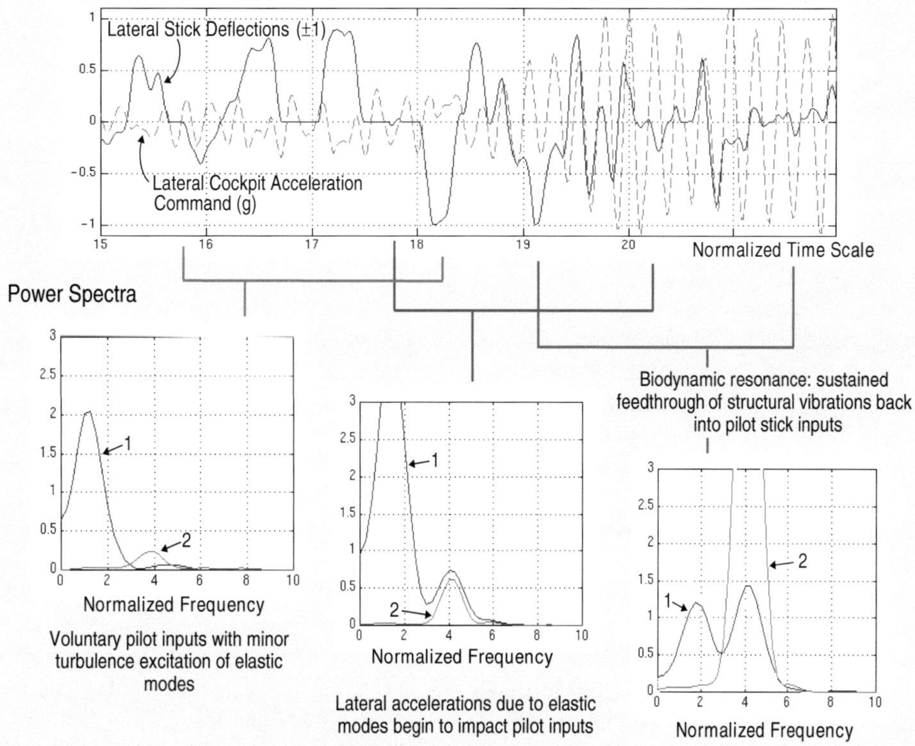

The upper graph shows the time histories of lateral cockpit acceleration and pilot lateral stick input for a moving-base piloted simulation of a demanding landing-approach task. These time histories reveal an undamped oscillatory instability beginning at about 19 normalized time units. The cause of this instability is revealed in the power spectra of these two time histories, shown in the three lower graphs, where curves labeled "1" show lateral stick power spectral density and curves labeled "2" show lateral acceleration power spectral density. In the first power-spectra plot, on the left, calculated early in the time history, there is little correlation between the peak-power frequency of the stick input and the lateral accelerations. But in the power-spectra plot on the right, which is calculated from the traces in which the instability is evident, the peak power of the stick input shows a strong correlation with the peak power of the lateral acceleration. This indicates a feedback process is present between the cockpit accelerations and the stick input. That is, the accelerations are driving the stick inputs, through the dynamic responses of the pilot's body to the stick, and these stick inputs are in turn driving the accelerations. These data further indicate that the instability is inadvertent—the pilot could not avoid the instability despite—or because of—stick inputs intended to maintain stable flight.

SOURCE: Raney et al. (2001)

handling- and ride-quality requirements and to design and verify the performance of control systems. The ability of national simulation facilities to deliver high-fidelity manned simulations of highly flexible aircraft may need to be upgraded.

HIGH-FIDELITY INTEGRATED DESIGN TOOLS

Highly integrated designs are required for virtually all new aerospace systems. For aircraft, the importance of integration increases with flight speed. Supersonic aircraft are much more sensitive to how components and disciplines are combined than subsonic aircraft, particularly with regard to APSE effects and sonic boom, as already discussed. Stringent requirements for component performance (with attendant development, manufacturing, cost, and operational issues), coupled with the economic and environmental challenges faced by commercial supersonic aircraft, leave little room for inefficiencies in the design of the airframe, engine, flight controls system, or other performance-critical systems.

Design integration tools should allow design teams to interact in the design of complex systems where technical and other factors (including cost) can be appropriately traded; to compress the design cycle time by concurrently considering all critical constraints and disciplines; to adapt quickly to changes in design and manufacturing processes; to easily accept new and improved tools; and to provide databases with levels of complexity appropriate to each task.

Fortunately, a substantial national investment has been made in tools for integrated design, including system engineering methods, multidisciplinary optimization methods, detailed discipline methods and interfaces, and design-integration frameworks. The aircraft design and manufacturing industry is heavily committed to improving such tools. Universities and the government also have critical roles in advancing the state of the art in many of these areas. NASA and the Department of Defense have both made significant investments in the development of advanced integration environments and tools, although NASA's flagship program in this area, the Integrated Synthesis Environments Initiative, was recently cancelled.

Despite the progress that has been made, important work remains to be done. Existing tools cannot model some key technologies (e.g., active controls), nor do they have sufficient model validity in all important disciplinary areas. At the broad technical scale, it is extremely important to begin with a full understanding of the design objectives and constraints, such as payload, range, takeoff gross weight (TOGW), noise, sonic boom, and cost, and to identify all the critical disciplines. This will prevent suboptimization and the debilitating effects of discovering, too late, that a first-order design driver (such as APSE in the HSR Program) has not been fully appreciated or adequately addressed. Many existing design schemes do not fit well with integration/optimization algorithms, and user-friendly frameworks that accommodate such schemes are not available. Off-the-shelf software and interface mindsets are needed. Faster mechanisms for geometric modeling are required for improved efficiency at both the conceptual and detailed design levels. For design teams that might be geographically dispersed, mechanisms for sharing the geometric models are lacking. Indeed, the management and sharing of information and data are themselves first-order issues, as is reducing the time for each design cycle. Very large quantities of data must be transferred; presently, both modeling and data management are much too labor-intensive. Analytical design tools, such as computational fluid dynamics and finite element modeling, have been greatly improved, but they often take so long to run that they are impractical in an iterative design context, and they are not robust enough to be fully integrated into a design framework. Other important issues for the development of advanced integrated tools are associated with uncertainties: how they propagate through a design and how to develop calibration and validation processes that quantify them. Too often, even after an optimization calculation has been made, new designs can be evaluated only by comparing them against a previous baseline rather than by making an absolute measurement of expected performance and comparing it against a validation metric.

Major benefits could be realized from the development and effective use of advanced design and integration tools, but there are significant barriers to achieving these benefits. On the nontechnical front, NASA could help by creating a new culture of collaboration, which is required for the most effective utilization of university, government, and industry talent in the realm of integration tools. NASA also has the charter and opportunity to provide much of the technology that is needed to enable viable supersonic aircraft designs.

First, it will be vital to develop advanced, high-fidelity methodologies and tools for intradiscipline analysis in areas such as computational fluid dynamics and finite element modeling for structures. Interdisciplinary and multidisciplinary tools are also crucial for integrated design of complex systems and entire vehicles. *Particular attention should be given to (1) integrating the design of mechanical systems with the design of electrical systems and software development and (2) factors such as computational speed and robustness that will increase the utility of new tools in an integrated design context.* Speed should come naturally with advances in computer hardware capability, but algorithms must be tailored to take advantage of the massively parallel computing environment.

Second, to realize the potential for design improvements through advanced tools, substantial new efforts must be focused on the automation of integration and validation. Today, design processes can require weeks to a couple of months to set up and compute the aerodynamic, weight, stability and control, aeroelastic, and other performance characteristics resulting from a configuration change. Optimizing in any context takes a number of such cycles. While multidisciplinary optimization techniques can reduce the optimization time dramatically, the setup time for the basic configuration is still counted in weeks and the validation of designs resulting from multidisciplinary optimization techniques, at least in the usual context of experimental verification, is extremely difficult because of the highly integrated nature of the process (it typically involves the use of sophisticated analyses to check design calculations). There is, as a result, a great need for focused research on how to validate highly integrated design capabilities. Clever combinations of analytical, computational, and experimental approaches may be necessary. Where experimental approaches are needed, it is very likely that current ground test capabilities are inadequate. In any case, only after successful application of these newly developed validation methods will confidence be high enough to encourage the widespread use of the new tools.

Third, design and integration frameworks should be developed to allow teams of analysts and designers at different locations to come together and immerse themselves in a user-friendly design environment. Such frameworks must be able to achieve a wide range of design and integration objectives and accommodate discipline-specific analyses at varying levels of breadth and depth. Having a common framework to carry out the conceptual, preliminary, and detailed design phases is important, and approaches ranging from approximate to high-fidelity must be accommodated. Framework technologies should also facilitate the exchange and management of large databases. Design team members must have access to the same data, new data must be readily transferable and modifiable, and the entire process must take place in a secure environment with user-friendly methods for preventing unauthorized alteration of or access to information. This requires emphasis on system-user interfaces and data-management tools. Frameworks should also allow the inclusion of manufacturing and operations parameters, which are important to affordability.

NASA should work closely with engine and airframe manufacturers and other industries, government agencies, and universities engaged in the development of integrated design tools and methods to define the specific characteristics required of advanced technology collaboration infrastructures and to develop a comprehensive plan for meeting those needs. An expanded discussion of the above issues and a comprehensive set of recommendations for action is contained in *Design in the New Millennium* (NRC, 2000).

VARIABLE CYCLE ENGINES

A variable cycle engine would, in theory, allow the propulsion system to be optimized for different flight conditions (e.g., takeoff, supersonic cruise, subsonic cruise, and landing). A variable cycle engine is similar to a conventional mixed-flow turbofan, except that it has an additional secondary outer bypass duct to increase the overall bypass ratio and, thus, the air flow handling capability. The second bypass stream improves TSFC and improves fan surge control by allowing the fan to pass a maximum amount of air throughout a broader flight regime. Unlike conventional turbofans, a variable cycle engine varies the bypass ratio to optimize performance for different flight conditions. Reducing the bypass ratio during cruise improves fuel efficiency, whereas increasing it during takeoff and landing reduces community noise. Ejector nozzles would still be needed to mix ambient air with the jet exhaust to reduce noise enough to meet community noise standards. With a variable cycle engine, however, the nozzle can be smaller, which reduces aircraft weight and improves the economic viability of the design.

One of the leading design approaches for a variable cycle engine has a core-driven fan stage directly in front of the high-pressure compressor to supercharge both the core and inner bypass flow streams. The term "variable cycle engine" has come to apply narrowly to this type of engine. However, other engine cycles can also adjust their bypass ratio during operations and thus fall within the general class of "variable cycle engines." These include the fan-on blade ("Flade") cycle and the turbine bypass engine with an inlet flow valve (TBE/IFV) cycle, both of which were investigated by NASA's HSR Program (NASA, 2001). Initial studies at NASA indicated that the additional design complexity of variable cycle engines outweighed the benefits (Berton, 1992). Research has continued, but most of it has been proprietary. Continued development of engines with advanced cycles, such as the variable cycle engine, that are compatible with high cruise efficiency, low community noise, and small, lightweight nozzles could lead to important breakthroughs in the realization of commercial supersonic aircraft.

REFERENCES

Agrawal, S., and A.G. Powell. 1991. "Supersonic Boundary-Layer Stability Analysis of an Aircraft Wing." *Journal of Aircraft* 28 (November): 721-727.

Berton, J., W. Haller, P. Senick, S. Jones, and J. Seidel. 1992. Comparative Propulsion System Analysis for the High Speed Civil Transport. Cleveland, Ohio: National Aeronautics and Space Administration.

Boeing Commercial Airplane Group. 1989. High Speed Civil Transport Study, Summary, NASA Contractor Report 4234, September.

George, F. 2000. "The Supersonic Business Jet," *Business and Commercial Aviation* (July): 44-48.

Jones, R., and D. Cohen. 1960. High Speed Wing Theory. Princeton, N.J.: Princeton University Press.

Jones, R., 1991. "The Flying Wing Supersonic Transport." *Aeronautical Journal*, March.

NASA (National Aeronautics and Space Administration). 2001. Available online at <http://www.lerc.nasa.gov/WWW/HSR/PSI.html>.

NRC (National Research Council), Aeronautics and Space Engineering Board. 1994. Aeronautical Facilities: Assessing the National Plan for Aeronautical Ground Test Facilities. Available online at <http://www.nap.edu/catalog/9088.html>. Accessed on September 13, 2001.

NRC, Aeronautics and Space Engineering Board. 1997. U.S. Supersonic Commercial Aircraft. Available online at <http://www.nap.edu/catalog/5848.html>. Accessed on September 13, 2001.

NRC. 2000. Design in the New Millennium—Advanced Engineering Environments (Phase 2). Available online at <http://www.nap.edu/catalog/9876.html>. Accessed on September 13, 2001.

Raney, David L., E. Bruce Jackson, Carey S. Buttrill, and William M. Adams. 2001. The Impact of Structural Vibrations on Flying Qualities of a Supersonic Transport. AIAA Paper 201-4006, presented at the AIAA Flight Mechanics Guidance, Navigation, and Control Conference, Montreal, August.

Rech, J., and C. Leyman. 1980. "Concorde Aerodynamics and Associated Systems Development." In AIAA Case Studies in Aircraft Design. Reston, Va.: AIAA.

Rethorst, S., and M. Kantner, 1996. Underwing Traveling Wave / Feedback Control System, Vehicle Research Corp., PD 135, August.

Rising, J., and J. Vadyak. 1999. Shock Suppression Supersonic Aircraft. U.S. Patent 5934607, August.

Seebass, A. 1994. The Prospects for Commercial Transport at Supersonic Speeds, AIAA 94-0017, Sixth Biannual W.F. Durand Lecture, Arlington, Va.: May 1994.

Soloviev, V., V. Krivstov, A. Konchakov, and N. Malmuth. 1999. Mechanisms of Shock Wave Dispersion and Attenuation in Weakly Ionized Cold Discharge Plasmas. IVTAN Perspectives of MHD and Plasma Technologies in Proceeding of Aerospace Applications Conference, Moscow, Russia, March 24-25, 1999.

Tracy, R., I. Kroo, J. Chase, and J. Viken. 1995. Exploiting Natural Laminar Flow for Economical Supersonic Transport. SAE Paper 952040 SAE Aerotech '95. September 21.

White, E., and W. Saric, 2000. Application of Variable Leading-edge Roughness for Transition Control on Swept Wings. AIAA Paper No. 2000-0283.

Yoshida, K., T. Iwamiya, Y. Ueda, and H. Ishikawa. 2000. "Boundary Layer Transition Analysis of the Scaled Supersonic Experimental Airplane." Proceedings of an International Workshop on Numerical Simulation Technology for Design of Next Generation Supersonic Civil Transport, National Aerospace Laboratory, Tokyo, January.

4

Areas Needing Continued Technical Development

Chapter 3 describes the need for focused research initiatives to pursue breakthrough opportunities. A broad range of technical activity must be supported to ensure that the technologies discussed in this chapter are mature enough to convince industry to develop them and meet designer's needs at the time a supersonic transport is laid down. Technology requirements include reducing propulsion noise and controlling emissions of carbon dioxide, various oxides of nitrogen, and, possibly, water vapor (at very high altitudes); structures and materials that can withstand the more hostile environment resulting from supersonic flight; cockpit displays and controls compatible with the special design features and operational requirements of commercial supersonic aircraft; and systems and procedures to address safety issues associated with cabin depressurization and exposure to ionizing radiation during high-altitude flights. Although advances in the above technologies may not be necessary to develop an SBJ, they are critical to the ultimate goal of developing a large commercial supersonic transport. Each of these topics is discussed in more detail below.

PROPULSION EMISSIONS AND NOISE

Significant advances in propulsion technologies over the levels currently available are needed to meet environmental limits for NO_x, CO_2, water vapor, and airport noise. At the same time, the propulsion systems must be lightweight and demonstrate both high propulsive efficiency and low fuel consumption if the supersonic transport is to have enough range and payload to be an economic success.

Gas Turbine Engines

Key propulsion system parameters include TSFC (thrust-specific fuel consumption), thrust-to-weight ratio, engine life, noise, and emissions. As shown in Table 2-1, goals for these parameters are easier to achieve for aircraft with lower cruise speeds and smaller size. The committee believes the propulsion system technology available today is sufficiently advanced to allow industry to develop the propulsion system for an SBJ with a Mach number of about 1.6. On the other hand, achieving the goals for TSFC and thrust-to-weight ratio for a Mach 2.4 HSCT would require greater propulsive efficiency and component performance, including an engine nozzle that weighs 75 percent less than current, state-of-the-art designs. Some type of variable cycle engine, as described in Chapter 3, is likely to be needed if most of the propulsion system performance goals are to be met.

Emissions

From an emissions point of view, supersonic aircraft differ from subsonic aircraft in three respects. First, a supersonic engine will generally operate at higher temperatures. Second, even if the propulsion goals for efficiency and emissions index are achieved (i.e., about 800 lb of fuel per passenger for supersonic flights of 5,000 to 6,000 NM compared with 500 lb of fuel per passenger for a subsonic flight of equal distance), a supersonic aircraft will create more combustion products than a subsonic aircraft. Third, the most fuel-efficient cruise altitude increases with speed. Supersonic aircraft will therefore cruise at higher altitudes than subsonic aircraft, and the products of combustion will be emitted into a different region of the atmosphere, with detrimental consequences, as described below.

While all of the differences become more significant as flight speed is increased, the most important issue is probably flight altitude. The altitude of the tropopause, which separates the troposphere from the stratosphere, changes with the season and latitude, varying between about 30,000 ft at the poles and about 60,000 ft at the equator. Most subsonic flight takes place in the troposphere (at altitudes below about 40,000 ft), whereas most supersonic flight takes place in the stratosphere. A commercial supersonic aircraft optimized to fly at Mach 1.6 will operate at an altitude of about 50,000 ft, whereas a transport designed to fly at Mach 2.4 would ideally operate at about 60,000 ft.

Three issues associated with engine emissions are particularly important for a commercial supersonic aircraft: stratospheric ozone, climate change, and local air quality.

Stratospheric Ozone

Atmospheric ozone concentration is shown in Figure 4-1, along with typical cruise altitudes for supersonic and subsonic aircraft. The concentration of ozone peaks at about 70,000 ft. This stratospheric ozone layer is important in that it absorbs ultraviolet (UV) radiation that can lead to skin cancer. Aircraft emissions that affect atmospheric ozone concentrations include NO_x, water, and particles or particle precursors. Because of differences in atmospheric chemistry at different flight altitudes, emissions from subsonic aircraft tend to increase ozone concentrations, while those from supersonic aircraft reduce ozone.

Concerns about the effects of NO_x emissions on ozone depletion have intersected with plans for the development of supersonic aircraft since the U.S. SST program in the 1960s and early 1970s. NO_x takes part in catalytic chemical reactions that destroy ozone in the atmosphere. The IPCC Special Report on Aviation (IPCC, 1999) compared two scenarios for UV radiation in the year 2050. The first assumed an all-subsonic fleet of commercial airlines. The second assumed that a fleet of 1,000 Mach 2.4 HSCTs would replace 11 percent of the subsonic fleet. The net effect of the supersonic fleet is predicted to be a reduction in column ozone of 1.3 percent. The health impacts of this reduction were estimated by the Environmental Protection Agency: a total of 7,100 deaths in the United States among people born between 1980 and 2000. The uncertainty for this study was substantial, with a range of –5,000 to +36,000 (EPA, 2001).

Recent studies indicate that the effect of NO_x is smaller than had been thought in the 1970s, and the use of combustors with an emissions index of about 5 (i.e., 5 g of NO_x produced for every kilogram of fuel burned) may be sufficient to keep NO_x within acceptable limits. Other studies, however, have raised another concern: Emitting water into the stratosphere destroys ozone because it affects the composition, growth, and reactivity of aerosol reactions and provides a source of HO_x radicals that enhance ozone loss (Kawa et al., 1999). This is a serious complication, because water vapor is an essential combustion product of hydrocarbon fuels. Furthermore, the stratosphere naturally contains very

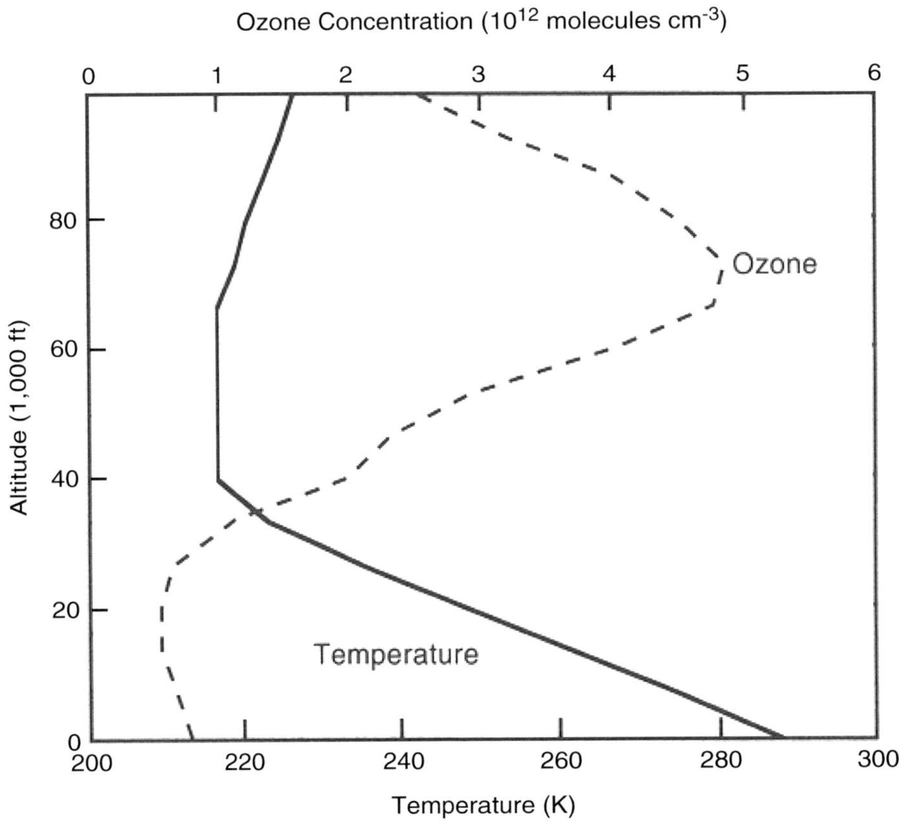

FIGURE 4-1 Schematic of the atmosphere between the ground and an altitude of 100,000 ft. SOURCE: Kawa et al. (1999).

little water and the rate of atmospheric exchange between the troposphere and stratosphere is very low. Thus, water emitted by aircraft into the stratosphere can have significant, long-lasting effects on the environment.

The effect of emissions from supersonic aircraft flying at a given altitude and route structure will be roughly proportional to the amount of fuel burned. If the size or number of supersonic aircraft is smaller than assumed in the IPCC report, the effect on column ozone would be less. The depletion of stratospheric ozone due to emissions from a fleet of supersonic aircraft could also be mitigated by cruising at a lower altitude. Extrapolation of the data plotted in Figure 4-2 suggests that ozone depletion could be eliminated by reducing average cruise altitude from 60,000 ft to about 50,000 ft. However, the uncertainty of results produced by current atmospheric models is still substantial. Continued development of these models is crucial to ensure that the environmental impacts of a future fleet of commercial supersonic aircraft can be accurately predicted.

Achieving an emissions index of 5 represents a reduction of 70 to 90 percent relative to current engines operating at similar conditions. NASA's HSR Program conducted small-scale steady-state and transient tests of combustor rigs to demonstrate that emissions indexes below 5 g/kg could be achieved. However, the program was terminated before ground and flight tests of an integrated engine could demonstrate the performance of the low emissions combustor technology in a flight engine that could also meet other essential performance parameters, such as thrust-to-weight ratio, TSFC, and reliability.

Another option for mitigating the effect of supersonic aircraft emissions on ozone would be to reduce the amount of sulfur in jet fuel. This would reduce the production of aerosol particles and increase fuel costs. The IPCC report assumed no change in the current jet fuel sulfur content, which is 0.05 to 0.07 percent by weight.

Climate Change

Climate changes can occur as a result of radiative forcing (i.e., changes in the global balance between incoming solar radiation and outgoing infrared radiation). Figure 4-3 shows the estimated impact on radiative forcing of a mixed commercial fleet of supersonic and subsonic aircraft in the year 2050, and Figure 4-4 compares the radiative forcing in 1992 with the radiative forcing in 2050 for both an all subsonic fleet and a mixed subsonic-supersonic fleet. With the mixed supersonic-subsonic fleet, radiative forcing from aircraft is

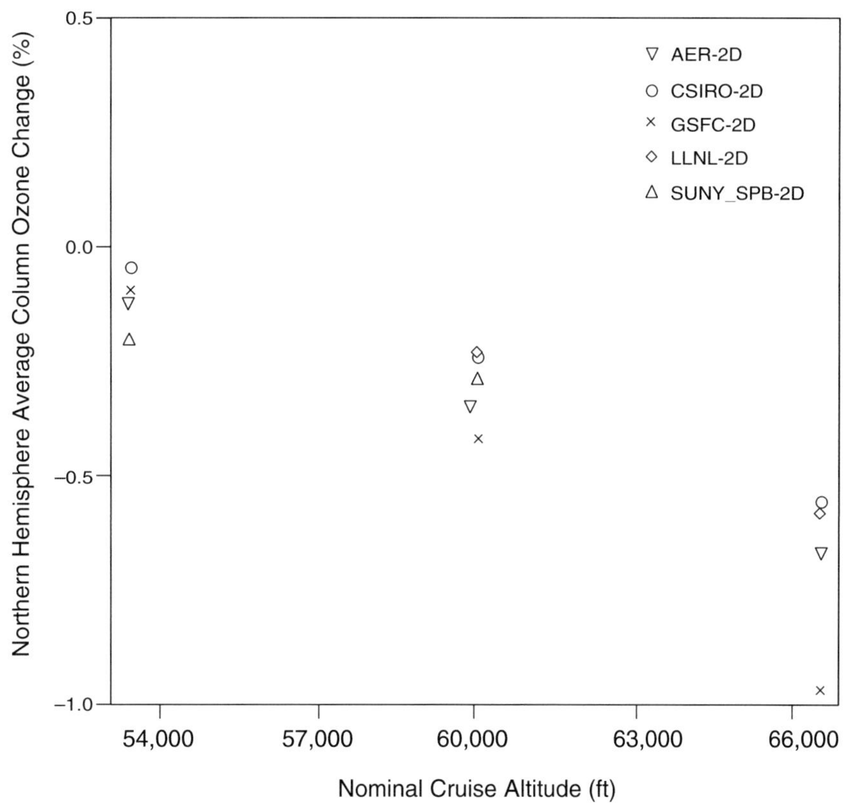

FIGURE 4-2 Sensitivity of predicted ozone change to a shift in cruise altitude (for a nominal cruise altitude of 60,000 ft). SOURCE: Kawa et al. (1999).

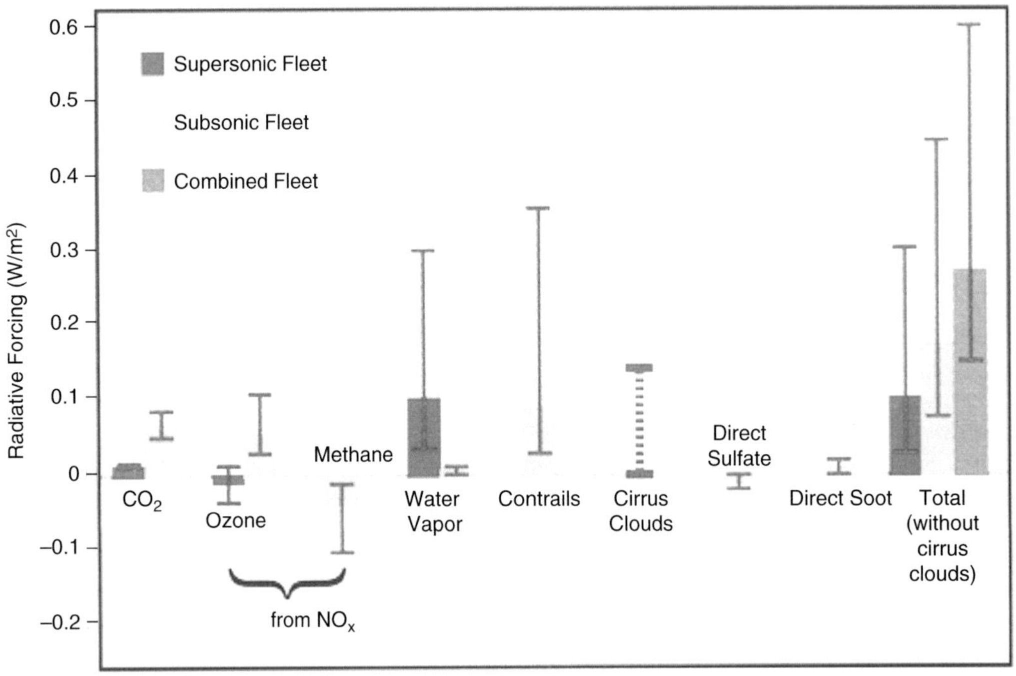

FIGURE 4-3 Estimate of globally and annually averaged radiative forcing for a worldwide commercial fleet in 2050 that includes subsonic and supersonic aircraft. SOURCE: IPCC (1999).

estimated to be 42 percent greater than with an all-subsonic fleet, because each supersonic aircraft at 62,000 ft increases radiative forcing by a factor of about five compared to equivalent subsonic aircraft flying at lower altitudes. However, as shown in Figures 4-3 and 4-4, the uncertainty associated with the estimates is very large.[1] The estimated radiative forcing effect of the mixed fleet in 2050 ranges from about 0.15 to 0.6 watts per square meter. Reducing this large uncertainty is essential to developing a better understanding of the environmental challenges posed by commercial supersonic aircraft.

The disproportionate effect of the supersonic transport fleet on radiative forcing is primarily due to larger emissions of water in the stratosphere and the lack of vertical mixing between the stratosphere and troposphere. As indicated in Figure 4-3, there is a large uncertainty (by a factor of about 3) surrounding the climatological impact of radiative forcing from aircraft water emissions. Better atmospheric models are needed to assess potential climate changes resulting from supersonic civil transports.

If the effects of water emissions in the stratosphere prove to be unacceptable, a technological breakthrough would be required. Some options to eliminate water emissions include the following:

- Emit water in a form that settles into the troposphere (e.g., as large droplets or ice particles).
- Remove and collect water from emissions and store it until it can be released into the troposphere.
- Use a fuel that does not contain hydrogen so that no water is formed.
- Generate thrust without chemical reactions onboard the aircraft.

The committee did not identify any efforts to develop any of these approaches for supersonic aircraft applications, and even with a focused effort a practical system is unlikely to be demonstrated for decades. Alternative propulsion technologies are discussed in more detail below.

Local Air Quality

The primary concern with aviation's effect on local air quality is the emission of NO_x and unburned hydrocarbons

[1] For the scenarios depicted in Figures 4-3 and 4-4, the bars indicate the best estimate of forcing, while the line associated with each bar depicts a two-thirds uncertainty range, meaning there is one chance in three that the true value falls outside the ranges shown. Available information on cirrus clouds was judged to be insufficient to determine either a best estimate or an uncertainty range; the vertical line indicates a range of possible best estimates. For the supersonic scenarios in 2050, supersonic aircraft are assumed to replace part of the subsonic fleet, reducing emissions from subsonic aircraft by 11 percent (IPCC, 1999).

near the ground (below 3,000 ft), because both can lead to ozone formation. (Near the ground, ozone is viewed as a pollutant because it is harmful to inhale and contributes to the formation of smog.) Assuming that future supersonic aircraft will not use afterburners (which are very fuel inefficient), a key difference between a supersonic aircraft and a subsonic aircraft with the same payload is that the supersonic aircraft will burn more fuel because aircraft designed for high-altitude supersonic flight are generally much less efficient during low-speed operations near the ground. Future supersonic aircraft are unlikely to receive exemptions from the environmental standards imposed on subsonic aircraft, so the need to minimize the impact on local air quality provides another reason to develop advanced supersonic propulsion technology.

Noise

Community noise is dominated by aircraft operations during takeoff, climb, and landing approach. It can be reduced by improving (1) propulsion systems, (2) the aerodynamic design of aircraft, and (3) the manner in which aircraft are operated.

Community noise from supersonic aircraft results primarily from mixing of the high-velocity jet exhaust with ambient air. Noise from the rotating blades of fans and compressors radiates from the inlet and contributes to the overall engine noise signature. Federal and local rules limit noise and, in some cases, restrict or prohibit operations of certain aircraft because of the noise they make. If future supersonic aircraft are required to meet the same standards as subsonic aircraft, they may need to be at least 3 to 4 dB quieter than current Stage 3 standards during takeoff, climb, and approach to landing.

At cruise Mach numbers of about 1.2 to 2.0, engine cycles with high bypass ratios (1.5 to 3.0), including variable cycle engines, can be used to reduce community noise while still meeting thrust requirements during takeoff and climb and propulsion efficiency requirements during cruise. This becomes more difficult for aircraft with cruise Mach numbers above 2.0; cruise efficiency requirements mandate the use of lower bypass ratios and, thus, higher jet exhaust velocities.

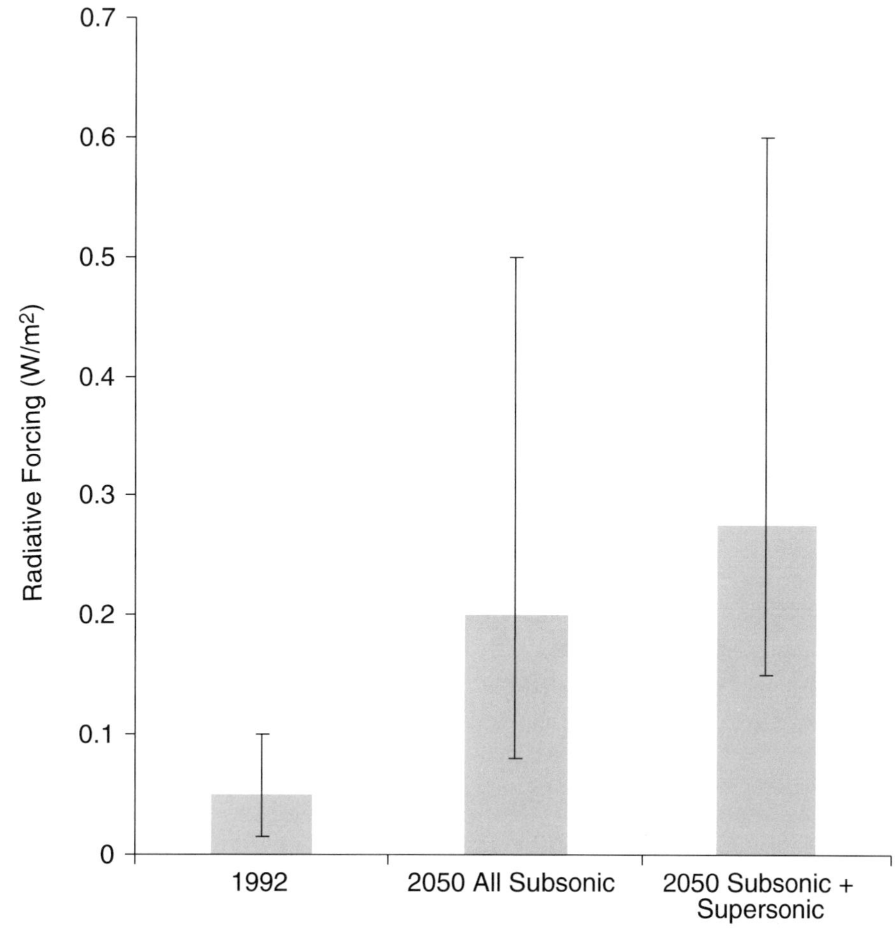

FIGURE 4-4 Comparison of estimated radiative forcing from the global fleet of commercial aircraft for (1) 1992, (2) 2050 with no commercial supersonic aircraft, and (3) 2050 with a fleet of large commercial supersonic aircraft. SOURCE: IPCC (1999).

Meeting noise standards then becomes increasingly difficult without a nozzle that is too large and heavy to be economical.

Other alternatives for meeting noise standards include the use of inlet choking to reduce inlet-fan noise on landing approach, injection of fluids into the exhaust, and active controls. NASA should consider expanding subscale acoustic tests on these alternatives, along with experimental activities on acoustic liners for exhaust ejectors.

Meeting noise standards will also require careful attention to operational procedures and low-speed aerodynamic characteristics. For example, noise could be reduced by changing Federal Aviation Regulations to allow the use of advanced/automated takeoff and climb procedures, including automatic variation of throttle setting. One- or two-segment, steep-decelerating landing approaches could also be used to reduce approach noise.

Aircraft designs with low aspect ratios and low wing loading are compatible with steep approaches to landing and have a large ground effect when the airplane is about a wingspan from the ground, increasing lift by about 20 percent. New, more complicated flight procedures will only be allowed, however, if flight control systems can handle the additional complexity without compromising flight safety.

The normal glide path for commercial transports preparing to land is between 2.5 and 3 degrees. A glide slope of 6 degrees or more is possible now at airports equipped with microwave landing systems. The steeper approach is quieter because it allows (indeed, it requires) reducing engine power, but it also requires the aircraft to stay within a much tighter approach envelope and creates a more difficult vertical navigation task. One solution would be to develop a steep approach guidance and control algorithm (with capabilities similar to the approach and landing mode energy management display on U.S. space shuttles). Such a system would (1) monitor the approach angle to make sure it is not too steep to safely land the aircraft on-speed and within the designated touchdown zone and (2) provide guidance to the pilot or autopilot based on aircraft energy state, configuration, and power settings.

Alternative Propulsion Technologies

In the long-term it may be possible to develop new propulsion systems that can be adapted to improve the performance of commercial supersonic aircraft or lessen their environmental impacts. Alternative propulsion technologies include pulse detonation engines, hydrogen-fueled engines, and fuel cell-based propulsion systems.

Pulse detonation engines (PDEs) use detonation waves to initiate rapid combustion of a pre-mixed fuel-oxidizer mixture contained in an array of tubes that are open at one end and closed at the other. Over the past 5 years industry and government laboratories, including NASA, have studied PDEs in configurations and test conditions consistent with aeronautical applications. For PDEs to become practical, however, internal losses must be reduced, the frequency of operation must be increased (to improve overall efficiency), and environmental performance must be improved. Detonation combustors tend to operate with a stoichiometric fuel-to-air ratio to increase thermal efficiency, but this may produce an unacceptable level of harmful emissions. Experimental PDEs are also intrinsically loud. Nevertheless, both the aircraft engine industry and the USAF are studying PDEs as augmenters for turbine-based engines (so-called hybrid PDEs).

Hydrogen-fueled turbine engines are attractive in that (1) modifying jet turbines to burn hydrogen is not technically difficult (the USAF did it in 1970 with a T-39) and (2) the exhaust product is mostly water vapor and NO_x. (Even though hydrogen fuel would contain no nitrogen, NO_x would be created by the high temperatures of the combustion process, which combines some of the oxygen and nitrogen in the air.) A key challenge in developing hydrogen-fueled engines is finding a compact way to store hydrogen. The volume required to store hydrogen in a gaseous form at reasonable pressures is so large that the slender shape of supersonic aircraft presents a serious structural challenge. Even in liquid form, the energy density of hydrogen by volume is one-fourth that of conventional aviation fuels. Also, although the energy density of hydrogen by weight is nearly three times as great as that of conventional aviation fuels, potential weight savings are offset by the additional weight of the high-pressure gas or liquid cryogenic systems needed to handle the hydrogen fuel and the additional aircraft structure needed to accommodate the large hydrogen fuel tanks and fuel handling systems. Two new means of storing high-density hydrogen may turn out to be useful as aviation fuel: slush hydrogen and gelled liquid hydrogen. The latter may have metallic additives as well, though such additives may create emission problems. The density improvement offered by these forms of hydrogen, as well as their advantages and disadvantages for aircraft application, is not yet clear.

In the early 1970s a NASA-funded study by Lockheed examined the potential of hydrogen-fueled supersonic aircraft with cruise speeds of Mach 2.2 and 2.7. The aerodynamic weight and propulsion characteristics of a previously established design for a Mach 2.7 supersonic cruise vehicle fueled by liquid hydrogen were critically reviewed and updated. NASA explored the effect of fuel price and noise restrictions on vehicle design and operating performance. The study concluded that, compared with equivalent aircraft powered by conventional jet fuel, aircraft fueled by liquid hydrogen offered potential advantages in performance, cost, noise, pollution, sonic boom intensity, and energy utilization (Brewer and Morris, 1975). However, the idea was apparently set aside. In 1997 Boeing indicated that it had preliminary designs for a hydrogen-fueled airplane, but it has no plans for pursuing this technology. The European Union, however, is supporting a partnership of 35 organizations coordinated by Airbus that is carrying out a 2-year systems analysis of subsonic commercial aircraft that could be fueled by liquid hydrogen. This project will examine the technical feasibility, safety, environmental compatibility, and economic viability of using liquid hydrogen as an aviation fuel for a wide variety of subsonic commercial aircraft, from business jets to very large, long-range commercial transports. Environmental concerns that bear examination include the effect of water vapor emissions (including contrails) and the environmental impacts of a hydrogen fuel production industry. The European effort, however, may finesse the latter concern by assuming that the production of hydrogen will be powered by renewable energy sources (EADS, 2001). A key economic issue is the potentially prohibitive cost of building cryogenic systems at airports and elsewhere for transporting and storing liquid hydrogen, combined with the long-term costs of maintaining and operating two fuel systems—one for hydrocarbon and one for hydrogen.

Fuel cells could be used to produce power for an electrically driven propulsion system. The key challenge for a practical fuel cell propulsion system is increasing power density (in terms of weight and volume). Other issues include peak power demands, power management, and the type of propulsion devices. The power density of state-of-the-art fuel cells has increased by a factor of 7 during the past 5 or 6 years, and a fuel-cell based propulsion system having twice the power density of current fuel cells could become technically feasible for a small, general aviation aircraft. However, the energy density needed for a commercial supersonic aircraft is orders of magnitude higher and would probably require fuel cells with room-temperature superconductors. Also, a fuel-cell aircraft would probably be powered by hydrogen and so would need to address the challenges already discussed for that fuel. One advantage of fuel cells would be the potential to collect the water vapor in the emissions (if the elimination of water vapor becomes a requirement, which is conceivable for stratospheric aircraft). Although this capability is difficult to imagine in a system suitable for a commercial supersonic aircraft, it would be easier to capture water vapor from the emissions of a fuel cell than from the exhaust of a turbine engine.

The idea of using fuel cells as the prime mover assumes that substantial advances in electric propulsion will have been made by the time fuel cell technology becomes a practical energy exchange device. Currently, conversion of electric power to propulsive power generally involves electric motors that are too heavy for aircraft applications. Electrically driven space propulsion engines have also been developed, but they have power densities that are many orders of magnitude too low for high-speed aircraft applications. Other, more radical approaches can also be imagined, such as zero-emissions aircraft with nuclear propulsion systems or beamed power systems using ground- or space-based power-beaming stations. Nuclear propulsion systems are unattractive, however, because of the need for very heavy radiation shields and the threat of radioactive contamination, especially in case of an aircraft accident. A power-beaming concept also raises fundamental issues, such as how to ensure continuity of power to a target moving at hundreds of miles per hour over large distances with a reliability of 0.999999999 ("nine 9's"—the standard for safety-critical aviation systems).

The committee believes that the alternative propulsion concepts examined are generally worthy of basic research support because they could be useful in various power or propulsion applications. However, none is applicable only to supersonic flight, and successful application to commercial supersonic aircraft would almost certainly have to be preceded by success in other applications, such as ground-based power systems (fixed or mobile) and subsonic aircraft. Furthermore, the economic and environmental justification for developing these technologies for existing, large-scale applications (such as electric power plants, automobiles, or subsonic aircraft) is probably much stronger than it is for applications that have yet to demonstrate long-term economic success (such as supersonic aircraft).

In light of the above, the committee has concluded it would be inappropriate to use the limited resources available for development of commercial supersonic aircraft technology to support basic research in alternative power and propulsion systems that show no particular promise for or relevance to supersonic applications. Supersonic R&T should focus on basic and applied research projects with the potential for resolving the difficult problem of improving the environmental acceptability of gas turbine engines powered by conventional fuels.

PROPULSION MATERIALS

For the next generation of supersonic commercial aircraft to meet environmental and economic requirements, propulsion materials and coatings must be superior to those used in the current generation of subsonic aircraft and in the Concorde. The degree of improvement is dependent on aircraft performance specifications, such as range, cruising speed, and size—an SBJ cruising at Mach 1.6 will need considerably less improvement than an HSCT cruising at Mach 2 or faster. More rigorous environmental and occupational safety regulations are expected to eliminate hazardous materials such as lead, chromium, and cadmium from high-temperature bearings and rotating surfaces. This could also prove to be a critical issue. Improvements are also needed in four other critical areas:

- combustor liner materials and coatings to meet emission and durability requirements

- turbine airfoil alloys and thermal barrier coatings to meet performance and durability requirements
- high-temperature alloys for compressor and turbine disks to meet performance and durability requirements
- strong, lightweight, high-temperature materials that will enable the propulsion system to meet noise and weight requirements

Other materials and coatings could also be critical to the next generation of supersonic propulsion systems. These include compressor and turbine seals with better resistance to erosion. Stiffer, lower density fan-containment materials could also prove to be critical, especially for cruise speeds in excess of Mach 2.

Propulsion materials are a high-risk technology for HSCTs. To reduce the weight and improve the performance of supersonic propulsion systems, materials must have low density, high strength, and long life at high temperatures. Extending the combustor temperature to 3000 ºF or nozzle temperature to between 2300 ºF and 2400 ºF will be a very challenging materials problem.

Advanced materials such as polymers, intermetallics, metal matrix composites, and ceramic matrix composites (CMCs), coupled with innovative structural designs, could significantly reduce engine weight while improving engine aerothermodynamics. Furthermore, advanced materials have the potential to reduce aircraft TOGW, fuel consumption, emissions, and noise. New fibers and fiber coatings will also be required to withstand the extreme temperatures and long operational times of many engine components.

Titanium aluminide may be used to fabricate several critical components for engine nozzles. The potential weight savings of titanium aluminide over conventional superalloys make it a good candidate for high-temperature applications where high stiffness is required. To capitalize on the potential for this class of material, more research is required. Areas for improvement include low-cost material production, robust joining methods, and more comprehensive databases for materials properties . In addition, further development may allow PETI-5 (phenylethynyl-terminated imide), a high-temperature resin created by NASA during the HSR Program, to help satisfy the requirement for a high-temperature composite matrix resin and adhesive.

Lower manufacturing costs would help achieve the economics that commercial supersonic aircraft need to be successful. Isothermal rolling and forging, cold and hot spray forming, and laser powder deposition could reduce manufacturing costs substantially by reducing the buy-to-fly ratio for materials from 10-to-1 to 2-to-1, especially when combined with advanced modeling and computer simulations. Large, complex structural castings have long lead times and are expensive. A third of the cost is associated with inspection, repair, and rework. More modeling and computer simulation could reduce these costs. Casting with laser powder deposition to produce complex castings could lead to additional cost savings.

Combustor Liner Materials and Coatings

Current subsonic engines use state-of-the-art nickel-based superalloys with a thin ceramic coating for additional thermal protection. To meet the targeted NO_x emissions index of 5, the volume of film cooling air in the engine combustor must be greatly reduced, though the actual requirements are strongly dependent on the combustor design concept. A final combustor design is required to determine how much the cooling air must be reduced—and how much the operational temperature of the liner must be increased. In any case, materials that can withstand considerably higher temperatures than nickel-based superalloys will probably be needed for combustor materials exposed to very high temperatures for very long times.

NASA's HSR Program concluded that CMC was particularly promising as a combustor liner material. However, CMC is expensive and there is not yet an established manufacturing process to produce CMC liners. CMC is also brittle, which could create a durability problem. The Department of Energy has been supporting CMC development for industrial gas turbine applications, and it has achieved over 20,000 hours of operation at low temperatures (2200 ºF to 2400 ºF). This experience is generating an operational database that will reduce both the technical and financial risk of using CMCs in aircraft. Another promising but less well-developed liner material is a molybdenum-niobium-based alloy. Successful development of a liner using CMC or a molybdenum-niobium-based alloy would greatly advance efforts to satisfy emission requirements for future commercial supersonic aircraft.

Turbine Airfoil Alloys and Thermal Barrier Coatings

One of the key differences between supersonic engines and subsonic engines is the profile of temperature versus time during a typical flight. In a subsonic engine, critical engine components, such as the disks and the turbine airfoils, are at maximum temperature only during takeoff and climb, which typically last about 15 minutes. In a supersonic engine, critical engine components are at maximum temperature during cruise. Thus, a supersonic flight that lasts 4 hours subjects critical engine components to high temperatures 16 times longer than a subsonic flight of the same distance. Consequently, the creep and thermal mechanical fatigue properties of the airfoil alloys must be much more robust than the alloys used in subsonic engines, even if the maximum operating temperatures are the same. In fact, new supersonic engines are likely to have turbine inlet temperatures a couple of hundred degrees higher than current subsonic engines. The combustor exit temperature will probably be between 3000 ºF and 3500 ºF. A new thermal barrier coating is needed to provide 200 ºF to 300 ºF of thermal protection. This will require a ceramic coating with thermal conductivity 50 to 75 percent lower than that of current thermal barrier coatings. The HSR Program made some progress in this area, developing an airfoil alloy and thermal barrier coating to a TRL of 3 to 4. However, this is still an important risk item, and the risk increases as the cruise Mach number increases.

High-Temperature Alloys for Compressor and Turbine Disks

To satisfy the performance, durability, and affordability requirements of an HSCT, the required pressure ratio will probably push compressor outlet temperature 100 ºF to 300 ºF higher than current subsonic engines. In addition, as discussed above, the disks will need to withstand elevated temperatures much longer than the disks in subsonic engines. The longer operational times at the higher temperatures will require the new disk alloys to have improved creep and cyclic fatigue properties. The HSR Program developed a new disk alloy with a 1350 ºF limit and advanced it to a TRL of 3 to 4. However, no full-size disks were forged and no component tests were conducted to determine the ability of the new alloy to meet the required disk service life of 18,000 hours. The temperature limit needs to be increased by another 100 ºF to 150 ºF, forging and heat treatment procedures must be validated to ensure that the disks can be manufactured, and component tests must be conducted to establish disk life. The lower limit is probably adequate for SBJs, and the higher limit is essential for higher-Mach-number airplanes.

Materials to Meet Engine Noise and Weight Requirements

The extent to which a commercial supersonic aircraft will require new materials to meet engine noise suppression requirements is highly dependent on the type of aircraft and the engine design concept. The need for engine nozzles to suppress noise would be much less for a smaller aircraft or an engine design cycle with a high bypass ratio during takeoff and climb. The primary nozzle issue is achieving weight, cost, and size goals compatible with aircraft weight and performance requirements. Concerns about the durability of the components when exposed to high levels of acoustic energy and high temperatures are also well founded. The HSR Program developed large, thin-wall castings of nickel superalloys and titanium aluminide intermetallic materials along with CMC acoustic tiles to a TRL of about 4. Continued development is needed to fabricate a subscale nozzle with these materials and conduct component tests in an appropriately simulated environment.

AIRFRAME MATERIALS AND STRUCTURES

Structural Durability

Material systems selected for commercial supersonic aircraft could deteriorate in the thermomechanical environment expected in service. Some existing materials, such as graphite epoxy, may be suitable for cruise speeds below Mach 2, while others, such as graphite/bismaleimides, may be suitable up to Mach 2.2, but testing is needed to verify that they will perform adequately after lengthy exposure to the high temperatures of supersonic flight. The airframe must perform its intended function throughout its time in service to ensure flight safety and to meet customers' economic expectations. Much work has been done over the last 20 years or more in understanding mechanisms of deterioration due to temperature, oxidation, ultraviolet, and time for several composite material systems. Also, some aluminum-lithium alloys have been exposed to high temperatures for prolonged periods of time. Little is known, however, about the applicability of that research to other material systems that may have different deterioration mechanisms. In addition, some preliminary accelerated aging protocols have been developed for high-temperature material systems, but more data are needed to prove that the protocols are valid. Time is of the essence for providing these data. About 7 years are needed to conduct simulated lifetime testing for a vehicle designed for 60,000 hours of operation at design temperature (i.e., at cruise conditions). This makes it impractical to include complete lifetime testing in preliminary screening of candidate materials or the generation of design data. A long-term commitment is required to validate accelerated testing protocols.

Proof of Structure

Traditional aircraft design methods include numerous structural tests of varying complexity—including tests of complete airframes—to prove that structures are satisfactory from both a static and a durability perspective. Appropriate thermomechanical tests can be conducted on small test articles, but similar tests on complete airframes can be prohibitively expensive. In addition, airframe testing may have little value, because it is difficult to validate the accuracy of the simulated environment.

Nevertheless, the certification process—and good engineering practice—requires manufacturers to prove that the airframe will perform safely throughout its time in service. Airframe manufacturers have accumulated a wealth of knowledge from past test programs. Several thermomechanical tests were conducted on the Concorde airframe; the full-scale fatigue test proved to be particularly challenging because of problems with simulating the in-flight heat transfer rates. Testing will become even more complicated if composite material systems are widely used, because the relationship between deterioration under mechanical loads and deterioration under thermal loads is not well characterized. This makes it impossible to simulate thermal effects using mechanical loads.

Structural Flexibility

The fuselages of commercial supersonic aircraft will be long and slender. Designing an airframe stiff enough to avoid APSE problems and light enough to allow an economic payload will be difficult. Conventional aircraft materials and structures will not be able to provide adequate stiffness at an acceptable weight, especially for HSCTs. At a minimum, increased stiffness will be required at several locations on the airframe, such as outboard wing and forward fuselage. Some work has been done on aeroelastic tailoring but little on the scale that would be required on an HSCT. Also, work will need to be done to improve the producibility of large structures with unbalanced lay-ups and to characterize their structural performance.[1] Hybrid laminates and materials with high-stiffness fibers, such as boron or high-modulus graphite fibers, are available, but more work is required to fully understand issues related to their performance in service. New fibers may need to be developed to provide increased stiffness without loss of strength or toughness. Large structures using the proposed materials and lay-ups will need to be manufactured and tested to prove viability.

Lightweight Structural Design Concepts

To meet weight and operational performance targets, lightweight structural design is essential. Sandwich construction has some potential, but it also has some limitations; moreover, existing data regarding its ability to reduce structural weight are inconclusive. This type of construction has been widely used in secondary applications on large commercial transports but is not often used for primary structure. Research supported by the HSR Program revealed that conventional skin-stringer construction would weigh less in areas subject to high loads and in wing and empennage structures having shallow box depth. The design information available for sandwich structure subjected to high loads is incomplete, especially where through-thickness effects are significant.[2] Previous research revealed that these through-thickness effects can limit the design by, for example, requiring higher density cores to provide (1) adequate core shear strength and (2) sufficient tension strength between the face sheet and the core. This would increase weight, making sandwich construction less attractive.

Another significant challenge will be developing a means of damage prevention following the absorption of water. Past service experience on subsonic commercial transports shows that it will be difficult to prevent sandwich structure from absorbing water throughout the life of a fleet. The freeze-thaw-boil cycle associated with an airplane flying at subsonic speeds at altitude followed by acceleration to supersonic speeds will prove to be an exceptionally difficult problem. Indeed, the rudder of the Concorde was susceptible to water absorption, which caused in-flight separation of portions of the rudder. Water absorption, however, is not a new problem and has been experienced in a variety of airplanes. The current solution for new airplanes is to fill the voids with closed cell foam. While this slightly increases weight, it reduces the water absorption rate by several orders of magnitude.

In addition, design loads are limited by the location of structural joints. Choice of load levels must consider factors such as durability and bearing strength. More efficient designs of conventional structures as well as new structural concepts will increase the likelihood of meeting design weight targets.

Structural Materials

There are many potential structural materials for commercial supersonic aircraft with cruise speeds of less than Mach 2, but few have been tested for long periods at the temperatures that would be experienced while cruising at supersonic speeds. Advanced materials may be advantageous in some structures for commercial supersonic aircraft with cruise speeds below Mach 2.0, but they are essential for faster aircraft. Graphite epoxies and aluminum cannot sustain the temperatures expected at cruise speeds above Mach 2.0 for long periods of time, and titanium is too heavy to use as the primary structural material. PETI-5 may be able to serve as the basis for composite materials for aircraft with cruise speeds between Mach 2.0 and Mach 2.4. Improved titanium alloys have also been developed for key structural applications (e.g., highly loaded wing spars). The HSR Program investigated a promising hybrid laminate consisting of titanium foil and high-temperature composite materials. While progress was made on different product forms of the PETI-5 material, most had not reached an advanced state of maturity when the HSR Program was cancelled. Life testing on some materials was in progress but was halted when it was still many thousands of hours short of reaching the equivalent of one design service lifetime. Accelerated aging protocols were postulated but not proven. No significant work was done on alternative fibers such as higher modulus graphite fibers or boron fibers for improved stiffness. Additional work was needed on the hybrid laminate, especially

[1] Conventional composite structures usually have a plane of symmetry in the thickness direction. For example, in honeycomb panels, each face sheet most commonly has an identical thickness, and an identical sequence of plies in the composite. In the case of composite laminates, the lay-up of plies is symmetrical about the mid-thickness. Unbalanced lay-ups can be difficult to fabricate without unacceptable distortion.

[2] Traditionally, honeycomb structures have been evaluated as if they were two-dimensional, with a focus on the loads and stresses in the plane of the honeycomb panel. As the size of the part increases, loads and stresses in the thickness direction may become significant and can no longer be ignored.

on development of a robust surface preparation for the titanium foil. Much more work was also required to improve the cost and producibility of these materials for fabrication of airplane parts. Materials using boron fibers have proven to be especially difficult to fabricate, and loose boron fibers can present a health hazard during handling.

Structural Adhesives

Adhesives that will perform at high temperatures for long periods of time are required, especially for operation above Mach 2.0. However, improved materials for lower temperature applications may also be advantageous, because many of the structural concepts that have been investigated rely on structural bonding.

Adhesives that perform consistently at high temperatures for long periods of time are needed to attain aircraft weight goals and ensure vehicle safety. Adhesives developed under the HSR Program were based on resin systems similar to those used in high-temperature polyimides, such as PETI-5. Early indications were promising, and both film and paste adhesives were developed. Other types of adhesives, such as a reticulate film adhesive, also need to be developed. While some work has been done on aging of adhesives, more data are needed. Only preliminary information on suitable accelerated aging protocols is available for these materials.

Sealants

For sealants, the application of maximum concern is fuel tanks on aircraft with cruise speeds above Mach 2.0; currently available sealants are probably sufficient for lower speed applications. The key characteristic of a fuel tank sealant is its ability to cure effectively in a few hours in a confined space at temperatures that can be easily attained in a production environment. Fuel tank leaks result in excessive downtime for repair and dissatisfied customers, so long life is also important. Additional lifetime testing is required to characterize long-term performance of sealants in the operational environment of a commercial supersonic aircraft. Additional research is required to explore the deterioration mechanisms and rates of candidate materials. Accelerated aging techniques are also needed.

Some candidate materials have been developed for fillet seals (i.e., seals along the edges of parts and around fastener heads) in fuel tanks. Suitable materials have not yet been identified for fay surface seals (between overlapping or fitted parts), and the only ongoing research is a small, proprietary effort by Japanese industry and government.

An alternative would be to develop a different approach to sealing fuel tanks. For example, fuel could be stored in bladders inside each fuel tank, although this is probably not a viable approach for a new commercial supersonic aircraft because of increased weight and the difficulty of removing bladders for structural inspections and repairs.

Coatings

Coatings are used for both appearance and thermal management. Without advanced coatings, it may not be possible to maintain fuel temperatures below acceptable levels for aircraft with cruise speeds above Mach 2.0. At Mach 2.4, fuel temperatures are predicted to reach the limit of acceptability: A maximum fuel tank temperature around 150 °F was assumed for design studies in the HSR Program.

Coatings with the desired emissivity and absorptivity are essential. The coatings must maintain these characteristics and remain aesthetically pleasing until the aircraft is repainted. The HSR Program viewed the development of advanced coatings as a low priority and did not fund research in this area. A small, proprietary, industry-funded study conducted concurrently with the HSR Program identified a few candidate materials (e.g., siloxane, fluoroethersilicone, and solgel with zinc ortho-titanate and titanium oxide pigments) that appeared to have many of the required characteristics, at least initially. Testing, however, indicated that the materials tend to turn yellow after only a few hundred hours at temperature. When the HSR Program was cancelled, the research on coatings was terminated before this problem could be solved. Additional resources are needed to more thoroughly investigate optical properties and to make sure new coatings can be applied easily and have adequate wear and erosion attributes.

COCKPIT SENSORS AND DISPLAYS

Supersonic flight imposes few unique requirements on avionics and flight deck systems. Government and industry are developing new avionics technology that will benefit commercial aircraft regardless of their range or speed. Indeed, advancements in avionics are often applicable to a wide range of commercial and military aircraft. The avionics and display technologies needed for a commercial supersonic aircraft are currently operational or in development at a TRL of at least 3.

New cockpit display technologies are of particular importance to the development of supersonic aircraft, because most supersonic aircraft designs feature a long, pointed nose for drag reduction. Additionally, most contemporary supersonic designs feature a modified delta wing optimized for high-speed flight that results in high angles of attack at lower speeds. The long nose and high angle of attack at low airspeeds impairs the forward visibility of the flight crew and requires the use of either a droop-nose design (like that of the Concorde) or advanced sensors and displays to restore visibility without the weight and mechanical complexity of the droop nose. Technology currently available or under development makes the second option feasible. For example, seamless merging of visual and radar imagery with information from a digital database has demonstrated a real-time, all-weather situational awareness display that provides the crew with terrain data, aircraft flight path information, weather, and traffic avoidance advisories. Three-dimensional imagery has intriguing possibilities and is actually being used for medical applications, but the current state of technology is not suitable for near-term airborne applications.

The development of these so-called enhanced vision systems is an interdisciplinary task that requires a wide spectrum of different information technologies: (1) reliable camera systems to provide visual imagery, (2) data link technology for transmission of guidance information, (3) complex databases to provide terrain data for synthetic images and object signatures to support the imaging sensors, (4) high-performance computer graphics systems to render synthetic images in real time, (5) onboard imaging sensors, such as solid state infrared or imaging radar, to provide a real view through darkness and adverse weather, (6) knowledge-based image interpreters to convert sensor images into a symbolic description, (7) projection technology for panoramic or holographic displays, and (8) precision navigation tools tied into the Global Positioning System or equivalent systems that may become available.

Advanced cockpit sensors and displays could enhance safety by detecting and displaying images of objects that pilots would not normally be able to see when looking through the cockpit window of a conventional aircraft. Safety could be further enhanced by integrating features such as navigation enhancements and proactive systems to avoid controlled flight into terrain and runway incursions. Flight crews, however, may be uneasy about flying aircraft with only a computer-generated view of the outside world. One way to deal with this uneasiness during approach and landing would be to locate the cockpit windows at the lower front fuselage of the aircraft, instead of the traditional location on the upper front fuselage. This would give the pilots direct visual contact during critical phases of the approach, landing, and ground operation. During other phases of flight (takeoff, climb, and cruise), pilots would use an enhanced vision system with computer-generated displays for safety, guidance, and control functions.

Other sensor and display technologies could further enhance the operability of commercial supersonic aircraft. Some systems, such as a boom shadow tracking system, would be unique to supersonic aircraft.[3] Supersonic aircraft would also benefit from upgrades to flight planning systems. Today, airline dispatchers and pilots consider weather and atmospheric conditions, especially winds and hazardous weather, to select safe flight routes that minimize flight time and fuel burn. Flight planning for supersonic flight operations would benefit from additional information, such as temperature anomalies and solar radiation. Routing could be further improved by increased use of real-time information to adjust routing in flight. Crews flying complex modern aircraft would also benefit from advanced cockpit concepts, such as a task saturation detection system. Conceptually, this system would detect stress and/or tunneled focus that might occur, for example, during abnormal or emergency flight conditions. If necessary, the system would assume temporary, but overrideable, control of certain flight functions to assure safety while allowing the crew to adequately handle the situation that captured their attention. This technology is currently at TRL 1 or 2.

IN-FLIGHT SAFETY

High-Altitude Radiation

Cosmic radiation is caused by high-energy charged particles from space (galactic cosmic radiation) and, to a lesser extent, from the Sun (solar cosmic radiation). Many of the particles are deflected by Earth's magnetic field. This effect is strongest at the equator and weakest at the poles; at altitudes commonly used by subsonic aircraft, cosmic radiation is twice as high at the poles as it is at the equator (FAA, 1990).

Cosmic-radiation particles that enter the atmosphere collide with atoms and molecules of air. This reduces the energy of the primary (incoming) radiation while producing secondary radiation in the form of lower energy particles and gamma rays. At sea level, Earth's atmosphere provides an effective shield against primary and secondary cosmic radiation, with a mass thickness of over 1,000 g/cm^2. The shielding mass drops off to about 200 to 300 g/cm^2 at normal cruise altitudes for subsonic aircraft (30,000 to 40,000 ft). At 65,000 ft, the shielding mass is 60 g/cm^2, which reduces the primary proton flux to about half of the incident flux, the alpha particle flux to about a quarter, and the heavy ion flux to about 3 percent or less, depending on the mass of the cosmic radiation particle (Goldhagen, 2000). When the Concorde entered service in 1976, cosmic radiation was identified as a potential crew safety issue. British Airways installed radiation-monitoring devices in the flight deck area and determined that radiation dosage received during flights at 50,000 to 60,000 ft were about twice the dosage received at 35,000 to 40,000 ft, the cruise altitude of a typical airliner.

On average, people living in the United States are exposed to about 3.5 millisievert (mSv) per year. Over 80 percent of this exposure is from natural sources, primarily radon and radon by-products in the atmosphere. At sea level, less than 10 percent of natural background radiation is from cosmic radiation (ISU, 1996). In an aircraft, however, primary and secondary cosmic radiation are the main source of radiation exposure. The average exposure of a commercial pilot on a single polar flight has been measured at about 0.1 mSv (Blakely, 2000). The FAA's advisory circular on radiation exposure (AC 120-52) estimates the annual exposure that crew members would experience if they worked full time

[3]A boom shadow tracking system would depict the location, coverage, predicted path, and strength of an aircraft's sonic boom based on standard, predicted, or real-time atmospheric conditions.

on one of 32 domestic and international routes. The largest annual doses, which were as high as 9.1 mSv, were for transoceanic flights at high altitude (up to 43,000 ft). By comparison, federal regulations require the nuclear power industry to ensure members of the general public receive no more than 1 mSv of additional radiation per year from nuclear power operations. The limit for radiation workers is 50 mSv per year (10 CFR 20).

AC 120-52 also addresses the health risk associated with the estimated exposures to cosmic radiation. If 1,000 crew members worked full-time (950 flight hours per year) for 20 years on a route with an annual exposure of 5 mSv, AC 120-52 estimates that about 6 might be expected to develop fatal cancers as a result. Based on public health statistics, about 220 of 1,000 Americans normally die from cancer, so exposure to 5 mSv annually would increase the chance of dying from cancer by 3 percent (FAA, 1990). The risk to passengers, who spend much less time in the air than full-time crew members, is much less.

Another early concern with high-altitude flight was the possibility of being exposed to excessive radiation as a result of solar flares. Conceivably, a strong solar flare could deliver a radiation dose greater than 10 mSv per hour to crew and passengers flying on polar routes. Concorde pilots have monitors so they can reduce altitude and latitude, if necessary. The Concorde routes are subpolar. Still, over the 25-year experience with the Concorde, no flight had to activate an emergency descent because of high radiation from a solar flare, even during peak years of the cycle.

The deployment of commercial supersonic aircraft should be accompanied by an update to AC 120-52 that extends the analyses of radiation exposure to the altitudes, routes, and flight times typical of supersonic aircraft, evaluates the need for tracking individual exposure by frequent flyers, and incorporates up-to-date knowledge regarding biological damage from highly energetic particles (which seems to be more severe than originally estimated in the 1960s) (Wilson, 2000). This update is necessary to ensure that aircrew and the flying public continue to enjoy a high level of confidence that high-altitude radiation does not threaten their health. Future research and analysis should assess the adequacy of current knowledge about high-altitude radiation with regard to the safe operation of commercial aircraft at various altitudes. Future research should also explore technological and operational approaches for reducing exposure to cosmic radiation, perhaps through lightweight shielding and/or the tracking of radiation doses accumulated by individuals and the exposure associated with particular flights (to make sure crew assignments do not result in excessive exposure to any individuals).

High-Altitude Cabin Depressurization

Cabin depressurization at altitudes higher than 50,000 ft would pose a safety risk to passengers and crew, especially in the case of a rapid or explosive decompression. Although the likelihood of a structural failure leading to cabin depressurization at cruise altitude is very small (failure of a compressor or turbine disk is the most likely cause), the effects could be fatal unless immediate and drastic action is taken. Above 50,000 ft, a person in good physical condition has a time of useful consciousness of only 5 to 12 seconds without supplemental oxygen. FAR 25.841 requires commercial aircraft to be designed so that occupants will not be exposed to a cabin pressure altitude that exceeds 25,000 ft for more than 2 minutes or a pressure altitude of 40,000 ft "for any duration" for any failure condition that cannot be shown to be "extremely improbable."

This could be a difficult standard to meet for commercial supersonic aircraft cruising at altitudes well in excess of 40,000 ft. Efforts to satisfy this requirement should investigate the innovative technologies such as self-sealing materials to contain fuselage pressure leaks. Even if self-healing technologies cannot seal leaks completely, they could reduce the leak rate and provide extra time for an emergency descent. An automatic emergency descent mode for aircraft flight control systems, which could be triggered by unexpected loss of cabin pressure changes, might also be necessary to meet safety standards.

CERTIFICATION

The FAA does not have certification requirements suitable for a future commercial supersonic transport. The Concorde was certificated using a set of special conditions. A preliminary set of rules, "Tentative Airworthiness Standards for Supersonic Transports," was developed during the U.S. SST program in the early 1970s. The FAA also established certification teams to support the future certification of new technologies under development by the HSR Program. Ongoing efforts to develop new supersonic technologies should proceed in parallel with the development of new regulatory standards to ensure that the regulatory approval process does not impede the development of a commercial supersonic aircraft. For example, certification issues could be a major obstacle to the use of a windowless cockpit that uses advanced sensor and display systems to produce a computer-generated view of the outside world. National resource specialists (i.e., FAA specialists in specified disciplines) could be a valuable resource for planning and implementing changes to certification standards.

REFERENCES

Blakely, E. 2000. "Biological Effects of Cosmic Radiation: Deterministic and Stochastic." *Health Physics: The Radiation Safety Journal* 79(5):495-506.

Brewer, G., and R. Morris. 1975. Minimum Energy Liquid Hydrogen Supersonic Cruise Vehicle Study. NASA CR137776. NASA Contract NAS2-8781 with Lockheed Martin, California. (N76-17101).

EADS (European Aeronautic Defence and Space Company). 2001. Special: Cryoplane. Available online at <http://www.eads-nv.com/eads/en/index.htm?/xml/en/businet/airbus/cryoplane/cryoplane.htm&airbus>. Accessed on September 13, 2001.

EPA (Environmental Protection Agency). 2001. Human Health Effects of Ozone Depletion from Stratospheric Aircraft. Washington, D.C.: EPA.

FAA (Federal Aviation Administration). 1990. Radiation Exposure of Air Carrier Crewmembers. FAA Advisory Circular 120-52, March 5. Available online at <http://ntl.bts.gov/DOCS/Ac12052.html>. Accessed on September 13, 2001.

Goldhagen, P. 2000. "Overview of Aircraft Radiation Exposure and Recent ER-2 Measurements." Health Physics: The Radiation Safety Journal 79(5):526-544.

IPCC (Intergovernmental Panel on Climate Change). 1999. IPCC Special Report on Aviation and the Global Atmosphere. Cambridge, U.K.: Cambridge University Press. Summary for policymakers available online at <http://www.ipcc.ch/pub/av(E).pdf>.

ISU (Idaho State University) Physics Department. 1996. Radiation and Risk. Available online at <http://www.physics.isu.edu/radinf/risk.htm>. Accessed on November 8, 2001.

Kawa, S., J. Anderson, S. Baughcum, C. Brock, W. Brune, R. Cohen, D. Kinnison, P. Newman, J. Rodriguez, R. Stolarski, D. Waugh, and S. Wolfsy. 1999. Assessment of the Effects of High-Speed Aircraft in the Stratosphere. Greenbelt, Md.: Goddard Space Flight Center.

Wilson, J. 2000. "Radiation Environments and Human Exposures." *Health Physics: The Radiation Safety Journal* 79 (5):470-494.

5

Findings, Conclusions, and Recommendations

For decades, the speed of commercial aviation was constrained by the sound barrier. Even with the Concorde, supersonic flight was unavailable except on a few routes and only for those willing and able to pay the high airfares. Commercially successful supersonic flight will only occur when technology is developed and assembled into an aircraft that can be profitably manufactured in large quantities (i.e., hundreds of aircraft) and that is affordable for users and environmentally acceptable to society at large.

Commercial supersonic aircraft could take many different forms. The path to success will depend upon the ultimate and intermediate goals selected and the time and resources set aside to achieve them. One approach would be to focus on entirely new forms of air vehicles, such as vehicles energized by laser beams from ground power stations. A second approach would be to pursue revolutionary new aircraft, such as a "morphing aircraft" with the ability to continuously modify the shape of its wings to optimize aerodynamic performance during all phases of flight. Either approach requires breakthrough technologies, and turning new technologies into an operational commercial supersonic aircraft is very expensive and takes decades of research and development to satisfy performance, economic, safety, and environmental requirements for aircraft and ground systems. Focusing NASA's efforts to develop technology related to commercial supersonic aircraft on long-term, high-risk concepts would probably result in a research program that does little or nothing to enable the operational deployment of environmentally acceptable, economically viable commercial supersonic aircraft in the next 25 years or less. Because that is the time frame of interest for this study, the committee endorsed a third approach: investing in breakthrough technologies that could be applied to the design of a more conventional commercial supersonic aircraft. The committee assessed the ability of advanced technology to meet customer and design requirements for the three types of commercial supersonic aircraft described in Table 5-1.

The committee concluded that foreseeable technological advances will be able to solve many key customer and design issues, particularly for commercial supersonic aircraft with cruise speeds of less than approximately Mach 2. But much work remains to be done. New, focused research is needed in several areas where existing efforts are unlikely to close the gap between the state of the art and aircraft requirements (Finding 1). However, new initiatives in these areas will be counterproductive if they divert resources from existing efforts that are also necessary to the development of a commercial supersonic aircraft (Finding 2).

TABLE 5-1 Commercial Supersonic Aircraft: Three Notional Vehicles

	Supersonic Business Jet	Supersonic Commercial Transport with Overland Capability	High-Speed Civil Transport
Speed (Mach no.)	1.6 to 1.8	1.8 to 2.2	2.0 to 2.4
Range (NM)	4,000 to 5,000	4,000 to 5,000	5,000 to 6,000
Payload (passengers)	8 to 15	100 to 200	300
Sonic boom low enough to permit supersonic cruise over land	Yes	Yes	Yes, if possible

Finding 1. An economically viable, environmentally acceptable supersonic commercial aircraft with a cruise speed of less than approximately Mach 2 requires continued development of technology on a broad front (see Finding 2). In addition, research in the following five areas of critical importance could lead to important breakthroughs, but only if current research is augmented by new, focused efforts (or significant expansions of existing efforts):

- airframe configurations to reduce sonic boom intensity, especially with regard to the formation of shaped waves and the human response to shaped waves (to allow developing an acceptable regulatory standard)
- improved aerodynamic performance, which can be achieved through laminar flow and advanced airframe configurations (both conventional and unconventional)
- techniques for predicting and controlling aero-propulsive servo-elastic and aircraft-pilot servo-elastic (APSE) characteristics, including high-authority flight- and structural-mode control systems for limiting both types of APSE effects in flight and tools for defining acceptable handling and ride qualities
- automated, high-fidelity, multidisciplinary optimization tools and methods for design, integration, analysis, and testing of a highly integrated, actively controlled airframe-propulsion system
- variable cycle engines for low thrust-specific fuel consumption, high thrust-to-weight ratio, and low noise

Finding 2. An economically viable, environmentally acceptable commercial supersonic aircraft with a cruise speed of less than approximately Mach 2 requires continued advances in many areas, particularly the following:

- airframe materials and structures for lower empty weight fractions and long life, including accelerated methods for collecting long-term aging data and the effects of scaling on the validity of thermo-mechanical tests
- engine materials for long life at high temperatures, including combustor liner materials and coatings, turbine airfoil alloys and coatings, high-temperature alloys for compressor and turbine disks, and turbine and compressor seals
- aerodynamic and propulsion systems with low noise during takeoff and landing
- cockpit displays that incorporate enhanced vision systems
- flight control systems and operational procedures for noise abatement during takeoff and landing
- certification standards that encompass all new technologies and operational procedures to be used with commercial supersonic aircraft
- approaches for mitigating safety hazards associated with cabin depressurization at altitudes above about 40,000 ft
- approaches for mitigating safety hazards that may be associated with long-term exposure to radiation at altitudes above about 45,000 ft (updating the Federal Aviation Administration's advisory circular on radiation exposure, AC 120-52, to address supersonic aircraft would be a worthwhile first step)

Conclusion 1. Research and technology development in the areas listed in Findings 1 and 2 could probably enable operational deployment of environmentally acceptable, economically viable commercial supersonic aircraft in 25 years or less—perhaps a lot less, with an aggressive technology development program for aircraft with cruise speeds less than approximately Mach 2.

The situation is somewhat different for commercial supersonic aircraft with cruise speeds above approximately Mach 2. The most efficient cruise altitudes are higher for higher cruise speeds. Aircraft with cruise speeds in excess of Mach 2 will normally cruise in the stratosphere, where engine emissions have a greater potential for climate change and depletion of atmospheric ozone. Water vapor, which is benign in the lower atmosphere, may have significant, long-lasting effects in the stratosphere and pose a serious limitation on high-speed, high-altitude flight. Also, at higher speeds, air friction creates higher temperatures. For cruise speeds of about Mach 2.4, new classes of structural materials are needed to meet strength and weight requirements. The noise suppression problem also becomes more challenging. The optimal engine bypass ratio is lower at higher Mach numbers, which leads to increased jet exhaust velocities and greater take-off noise. This makes it more difficult to design a nozzle that provides adequate noise suppression without becoming so large and heavy that the vehicle becomes uneconomical.

Finding 3. An economically viable supersonic commercial aircraft with a cruise speed in excess of approximately Mach 2 would require research and technology development in all of the areas cited in Findings 1 and 2. In addition, significant technology development would be needed to overcome the following barriers:

- climate effects and depletion of atmospheric ozone caused by emissions of water vapor and other combustion by-products in the stratosphere
- high temperatures experienced for extended periods of time by airframe materials, including resins, adhesives, coatings, and fuel tank sealants
- noise suppression at acceptable propulsion system weight

Conclusion 2. Candidate technologies for overcoming environmental barriers to commercial supersonic aircraft with a cruise speed in excess of approximately Mach 2 are unlikely to mature enough to enable operational deployment of an environmentally acceptable, economically viable Mach 2+ commercial supersonic aircraft during the next 25 years.

The environmental barriers associated with development of a supersonic transport with a cruise speed in excess of approximately Mach 2 are especially challenging. The most likely path to operational deployment of an environmentally acceptable, economically viable commercial supersonic aircraft within the next 25 years lies in research areas listed in Findings 1 and 2 as they would apply to an aircraft with a cruise speed of less than approximately Mach 2. Far-out technologies may be interesting and worth investigating, but they have little to do with development of a commercial supersonic aircraft in the foreseeable future.

Breakthrough aviation technologies are generally proprietary, competition-sensitive, and/or classified. Few individuals or organizations are willing to divulge publicly detailed information on such technologies. However, once the government has identified areas of interest and funds an appropriate research program, industry will respond with detailed proposals. For example, DARPA's Quiet Supersonic Platform Program is developing technology that would contribute to advanced supersonic aircraft with substantially reduced sonic boom, reduced takeoff and landing noise, and increased efficiency. Research areas include airframe shaping to reduce sonic boom, integration of low-specific-thrust propulsion systems, advanced inlet concepts, particulate injection into the engine exhaust, supersonic laminar flow, ceramics, and localized flow heating to increase virtual body length (and reduce sonic boom). Although detailed information about this research is still not publicly available, NASA has full access to it. In fact, the Quiet Supersonic Platform is managed by a NASA employee assigned to DARPA.

Most of the technology challenges identified by the committee are unique to supersonic aircraft. However, in some cases the technological advances necessary for development of commercial supersonic aircraft would also improve the performance of subsonic aircraft. Where applicable, efforts to develop supersonic technology should take advantage of related efforts to improve the performance of subsonic technologies and the lessons learned that come from operational use of new technologies in subsonic aircraft.

Recommendation 1. NASA should focus new initiatives in supersonic technology development in the areas identified in Finding 1 as they apply to aircraft with cruise speeds of less than approximately Mach 2. Such initiatives should be coordinated with similar efforts supported by other federal agencies (e.g., the DARPA Quiet Supersonic Platform Program).

Recommendation 2. For the technologies listed in Finding 2, NASA should allocate most of the available resources on goals and objectives relevant to aircraft with cruise speeds of less than approximately Mach 2. NASA should focus remaining resources on the areas listed in Finding 3 (i.e., the highest risk areas for cruise speeds greater than approximately Mach 2). Again, NASA activities should be coordinated with similar efforts supported by other federal agencies.

For each of the HSR Program's critical technology elements, the general goal was to demonstrate a TRL of 6 (i.e., system/subsystem model or prototype demonstrated in a relevant environment). This goal was appropriate, in part because of the large investment required and the high risk that individual technologies might not lead to a viable commercial product. In fact, even if the HSR Program had been completed, additional fundamental technology development and validation would have been required to prepare and demonstrate that critical technologies were ready for use in a commercial transport (NRC, 1997). Maintaining a goal of TRL 6 for supersonic research is essential if the results are going be adopted by commercial product development programs. NASA should continue to recognize TRL 6 as the appropriate goal for supersonic technology programs intending to transfer research results to a commercial product.

Recommendation 3. NASA and other federal agencies should advance the technologies listed in Findings 1 and 2 and Recommendations 1 and 2 to technology readiness level 6 to make it reasonably likely that they will lead to the development of a commercial product.

CLOSING REMARKS

Affordable, reliable, and safe air transportation is important to quality of life and economic growth. The global transportation infrastructure would be enhanced by the addition of a truly high-speed transportation element. If the United States intends to maintain its supremacy in the commercial aerospace sector, it has to take a long-term perspective and channel adequate resources into research and technology development. The technological challenges to commercial supersonic flight can be overcome, *as long as the development of key technologies is continued.* Without continued effort, however, an economically viable, environmentally acceptable, commercial supersonic aircraft is likely to languish.

REFERENCE

NRC (National Research Council), Aeronautics and Space Engineering Board. 1997. U.S. Supersonic Commercial Aircraft. Washington, D.C.: National Academy Press.

Appendixes

A

Biographies of Committee Members

DIANNE S. WILEY, chair, recently joined the Boeing Company Phantom Works, where she is program manager for transfer of advanced structure and materials technology to next-generation, reusable launch vehicles. Previously, she was with Northrop Grumman for 20 years, where she had been manager of Airframe Technology. In that position, Dr. Wiley was responsible for research and development and technology transition in structural design and analysis, materials and processes, and manufacturing technology. During that time, she was responsible for transitioning airframe core technologies into three new business areas (space, biomedicine, and surface ships) to offset declines in traditional business. Previously, as a senior technical specialist on the B-2 program, Dr. Wiley was responsible for developing and implementing innovative structural solutions to ensure the structural integrity of the B-2 aircraft. Dr. Wiley's 25 years of technical experience have involved durability and damage tolerance, advanced composites (organic and ceramic), high-temperature structures, smart structures, low-observable structures, concurrent engineering, and rapid prototyping. Dr. Wiley holds a Ph.D. in Applied Mechanics from the UCLA School of Engineering and Applied Science. She attended Defense Systems Management College in 1996, was a 1995 graduate of the Center for Creative Leadership, and was a member of the Leadership California Class of 1998.

H. LEE BEACH, JR., has expertise in aerospace technology research and development. Dr. Beach is a professor in the Department of Physics, Computer Science, and Engineering at Christopher Newport University (CNU). He is also an associate director (for CNU) at the Applied Research Center, a consortium of four universities and the Department of Energy, which conducts collaborative, applied research and technology transfer to local and regional high-technology businesses. In 1998, Dr. Beach retired from NASA as deputy director of the Langley Research Center. Previously, he was director for the National Aero-Space Plane Directorate at NASA Headquarters. During earlier assignments at Langley, Dr. Beach was deputy director for aeronautics and served as head of the Combustion Section, head of the Hypersonic Propulsion Branch, and acting chief of the High-Speed Aerodynamics Division. During his NASA career, Dr. Beach also led several studies to define the future direction of the nation's aeronautics programs and the size and makeup of the infrastructure to support them. For example, in 1993, he was codirector of the aeronautics portion of the National Facilities Study, which recommended consolidations and closures of some facilities, as well as the construction of two new national wind tunnels. The results of this study were validated by a concurrent NRC study.

JAMES A. (MICKY) BLACKWELL has expertise in airframe aerodynamics and manufacturing. His professional experience includes the development and manufacture of supersonic and subsonic aircraft, and he is familiar with technical issues associated with developing a supersonic business jet. Mr. Blackwell retired in February 2000 as executive vice president of the Lockheed Martin Corporation, where he had corporate oversight of the aeronautics business. Previously, he was president of Lockheed Aeronautical Systems Company in Marietta, Georgia; chief engineer for special projects; vice president of engineering; and vice president of the F-22 fighter program.

EUGENE E. COVERT, NAE, has expertise in aerospace technology research and development. Dr. Covert is the T. Wilson Professor of Aeronautics (Emeritus) at the Massachusetts Institute of Technology. He retired in 1996 after a long and distinguished career in aeronautics. Dr. Covert was associate director of the MIT Aerophysics Laboratory until he became the director of the Gas Turbine Laboratory and department head from 1985 to 1990. Dr. Covert has been both a member and chair of the U.S. Air Force Scientific

Advisory Board and the ASEB and has served on at least 10 NRC study committees. He is a member of the New York Academy of Science, a fellow of the Royal Aeronautical Society, and an honorary fellow of the American Institute of Aeronautics and Astronautics (AIAA).

DONALD M. DIX has expertise in propulsion systems. Dr. Dix retired from the Department of Defense (DoD) in May 1999 and is currently a consultant to government and industry. His last assignment with the DoD was as director for special programs within the Office of the Director of Defense Research and Engineering. Dr. Dix is a propulsion expert with broad experience formulating guidance and overseeing science and technology efforts for air platforms, space platforms, ground and sea vehicles, and materials and structures. He is also co-chair of the Independent Review Group that is supporting the Quiet Supersonic Platform Project and other work by the Defense Advanced Research Projects Agency (DARPA) to develop supersonic aircraft technologies. Dr. Dix has served on two NRC committees.

WILLARD DODDS has expertise in propulsion emissions technology and regulations. He is the senior staff engineer for emissions, regulations, and strategy at GE Aircraft Engines, one of two U.S. manufacturers of large jet engines. He is an expert in all aspects of aircraft engine combustion system design and development, including the design and development of low-emission combustion systems. As such, he has an expert knowledge of engine emissions abatement technology and relevant regulatory considerations. For the past several years, he has been the GE Aircraft Engines representative on various industry committees that interact with the International Civil Aviation Organization on engine emissions regulatory issues.

ILAN KROO has expertise in aircraft systems integration. Dr. Kroo is a professor of aeronautics and astronautics at Stanford University, where he conducts research in applied aerodynamics, aircraft design, and multidisciplinary optimization. Dr. Kroo has served as a member of one other NRC committee and for the last 6 years has been involved with DARPA's high-altitude, long-endurance aircraft programs. Dr. Kroo also has several years' experience with NASA Ames Research Center as a research scientist analyzing new aircraft concepts.

DOMENIC J. MAGLIERI has expertise in propulsion noise and sonic booms. He began his career at NASA (then NACA) Langley Research Center. He served as head of the Noise Control Branch of the Acoustics and Noise Reduction Division, responsible for developing technology for understanding, predicting, and applying solutions relative to noise generation, propagation, prediction, and reduction. Mr. Maglieri, who retired from NASA in 1986, works as a consultant for Eagle Aeronautics, Inc., where he is the lead engineer on all noise and sonic boom work for aircraft and space vehicles. He has almost 50 years of experience in airplane and helicopter noise, subsonic and supersonic transport technology, and sonic boom. He is considered a leading national and international expert on sonic booms. His sonic boom flight test involvement began in 1957 and has continued throughout four decades. Mr. Maglieri has participated in every major sonic boom flight test program involving 20 different aircraft, Apollo spacecraft, and the space shuttle. He has authored or coauthored over 150 publications, 95 of which pertain to sonic booms.

MATTHEW MILLER has expertise in airframe service life. As an employee of the British Aircraft Corporation, Dr. Miller worked on development of the Concorde supersonic transport, focusing on damage tolerance certification. Since joining Boeing in 1979, he has worked on the development and application of damage tolerance methods for all Boeing commercial aircraft products. He served as the structures manager of Boeing's high-speed civil transport program until that effort was cancelled in 1999. Dr. Miller is currently manager of Boeing's Structural Damage Technology organization.

DORA E. MUSIELAK has expertise in propulsion emissions and combustion. Dr. Musielak is currently engaged in studies of high-speed reacting flows—investigating injection, mixing, and ignition processes—as a member of the faculty of the Mechanical and Aerospace Engineering Department of the University of Texas at Arlington. Dr. Musielak has research, academic, and industry experience in aero and space propulsion, hypersonics, combustion, fuel injection and atomization for gas turbines, pulse detonation engines, magnetohydrodynamics, and development of space systems. Dr. Musielak conducted some of this research for Allison Gas Turbine, Solar Turbines, and the University of Tennessee Space Institute. Dr. Musielak is the recipient of two NASA research fellowships.

DAVID K. SCHMIDT has expertise in aircraft dynamics, stability, and control. He is a professor of mechanical and aerospace engineering, Director of the Flight Dynamics and Control Laboratory, and Dean of the Graduate School at the University of Colorado, Colorado Springs. Dr. Schmidt is also past chairman of AIAA's technical committee on guidance, navigation, and control and a past associate editor of the *Journal of Guidance, Control, and Dynamics*. Dr. Schmidt has served on one other NRC committee (the Committee on High Speed Research) and the U.S. Air Force Scientific Advisory Board's Review Panel for Science and Technology. He is a fellow of the AIAA.

MICHAEL WINSLOW has expertise in piloting and flight deck technologies. Mr. Winslow is an account team leader for Honeywell Defense & Space, with the responsibility for identifying and coordinating science and technology opportunities at Wright-Patterson Air Force Research Laboratory and NASA Glenn Research Center. Mr. Winslow is a former flight instructor with experience in military transports, military helicopters, and commercial transports. His professional experience includes 31 years as a pilot and flight commander of transport aircraft for the U.S. Air Force and the Air National Guard and a parallel career in the aerospace industry. Mr. Winslow has participated in the development of new aerospace products such as lighting systems, composite materials, and flight controls.

BILL G.W. YEE has expertise in propulsion materials. As director of materials and process engineering at Pratt & Whitney (now retired), Dr. Yee was responsible for managing and providing direction for the research, development, and characterization of materials and processes for application to commercial aircraft, military aircraft, and spacecraft propulsion systems and advanced components. He was also responsible for the development and maintenance of the material and process specifications and the approval and certification of suppliers and vendors. Dr. Yee also has 23 years of experience at General Dynamics, Fort Worth, where he was manager of advanced structures and design for aircraft systems. Dr. Yee has been a member of one NRC committee and one board (the National Materials Advisory Board).

Liaison from the Aeronautics and Space Engineering Board

ROBERT C. GOETZ, the liaison from the Aeronautics and Space Engineering Board to the study committee, is currently vice president of engineering at the Lockheed Martin Skunk Works. Prior to joining Lockheed, Mr. Goetz held a variety of positions during his 29-year career with the National Aeronautics and Space Administration, including deputy center director at the Lyndon B. Johnson Space Center, where he managed major space vehicle and space development programs. He conducted research in hypersonic aeroelasticity, was appointed head of the Flight Loads Section of the Structures and Dynamics Division and head of the Dynamic Loads Branch, and managed all research work in Materials, Structures and Dynamics, Loads and Aeroelasticity, and Acoustics and Noise Reduction at NASA Langley. He left NASA in 1987. Mr. Goetz is a fellow of the AIAA and the American Astronautical Society.

B

Participants in Committee Meetings

The full committee met four times from October 2000 through March 2001. As part of the committee's information-gathering process, smaller meetings were also attended by one or more committee members and representatives of public and private organizations. The committee wishes to express its thanks to invited guests and members of the public who participated in committee meetings and telephone conferences, including the following:

Noriaki Arakawa, Society of Japanese Aerospace Companies
Tom Auxier, Pratt & Whitney
Mike Bair, Boeing Corporation
Dick Bateman, Boeing Corporation
Mike Bauer, Executive Jet
Thomas Bauer, Vehicle Research Corporation
Sam Bruner, Raytheon
Dennis Bushnell, NASA Langley Research Center
Don Campbell, NASA Glenn Research Center
Dianne Chapman, NASA Glenn Research Center
Susan Cliff, NASA Ames Research Center
Jerold Creedon, NASA Langley Research Center
Ellis Cumberbatch, Claremont Graduate University
Dave Ercegovic, NASA Glenn Research Center
Bill Gilbert, NASA Langley Research Center
Ed Glasgow, Lockheed Martin
Pres Henne, Gulfstream Aerospace
Ray Hicks, NASA Ames Research Center
S. Michael Hudson, Rolls-Royce/Allison
Antony Jameson, Stanford University
Anjaneyulu Krothapalli, Florida State University
Hirotoshi Kubota, University of Tokyo
Brenda Kulfan, Boeing Corporation
Mary Jo Long-Davis, NASA Glenn Research Center
Joseph Luquire, MBL International, Ltd.
Robert Mack, NASA Langley Research Center
Harvey Maclin, GE Aircraft Engines
Koji Masuda, Japan Aircraft Development Corporation
Gordon McKenzie, United Airlines
Leik Myrabo, Rensselaer Polytechnic Institute
Tregenna Myrabo, Lightcraft Technologies
Chet Nelson, Boeing Corporation
Keith Numbers, Wright-Patterson Air Force Base
Robert Pearce, NASA headquarters
Peter Radloff, Boeing Corporation
Kenneth Reifsnider, Virginia Tech
Scott Rethorst, Vehicle Research Corporation
Michel Rigault, Dassault Aviation
John Roundhill, Boeing Corporation
John Seidel, NASA Glenn Research Center
Gary Seng, NASA Glenn Research Center
Robert (Joe) Shaw, NASA Glenn Research Center
Kevin Shepherd, NASA Langley Research Center
Rand Simberg, Vehicle Research Corporation
Richard Smith, Executive Jet
William Strack, NASA Glenn Research Center (retired)
Jack Suddreth, SRS, Inc.
Robert Tacina, NASA Glenn Research Center
Naohito Tsuda, Society of Japanese Aerospace Companies
Frank Tuck, Wright-Patterson Air Force Base
Olivier Villa, Dassault Aviation
Chowen Way, NASA Glenn Research Center
Alan Wilhite, NASA Langley Research Center
Richard Wlezien, Defense Advanced Research Projects Agency
Hidejiro Yamada, Society of Japanese Aerospace Companies
Tsutomu Yoshimura, Japan Aircraft Development Corporation

C

Acronyms and Abbreviations

AIAA	American Institute of Aeronautics and Astronautics
APSE	aircraft-pilot servo-elastic or aero-propulsive servo-elastic (phenomena)
CAEP	Civil Aviation Environmental Protection Committee (of the International Civil Aviation Organization)
CMC	ceramic matrix composites
CO_2	carbon dioxide
DARPA	Defense Advanced Research Projects Agency
FAA	Federal Aviation Administration
FAR	Federal Aviation Regulations
ft	feet
g	gram
hr	hour
HSCT	high-speed civil transport
HSR	High Speed Research (Program)
ICAO	International Civil Aviation Organization
kg	kilogram
lb	pound
L/D	lift-to-drag ratio
LERD	limited exclusive rights data
mSv	millisievert (a unit of ionizing radiation calibrated to measure potential biological harm; 1 mSv is roughly the daily dose due to natural background radiation)
NASA	National Aeronautics and Space Administration
NM	nautical miles
NO_x	oxides of nitrogen
NRC	National Research Council

PETI-5	phenylethynyl-terminated imide (a high temperature composite resin)
PDE	pulse detonation engine
psf	pounds per square foot
QSP	Quiet Supersonic Platform (Program)
R&T	research and technology
SBJ	supersonic business jet
SO_x	oxides of sulfur
SST	supersonic transport (either the program or the aircraft of the same name)
TOGW	takeoff gross weight
TRL	technology readiness level
TSFC	thrust-specific fuel consumption
TSFC/M	thrust-specific fuel consumption normalized to Mach number
USAF	U.S. Air Force
UV	ultraviolet